MAPMATICS

MAPMATICS

A Mathematician's Guide to Navigating the World

PAULINA ROWIŃSKA

THE BELKNAP PRESS OF
HARVARD UNIVERSITY PRESS
Cambridge, Massachusetts
2024

First published by Picador, an imprint of Pan Macmillan, a division of
Macmillan Publishers International, 2024

Printed in the United States of America
First Belknap Press of Harvard University Press edition, 2024

Library of Congress Cataloging-in-Publication Data

Names: Rowińska, Paulina, author.
Title: Mapmatics : a mathematician's guide to navigating the
world / Paulina Rowińska.
Description: Cambridge, Massachusetts : The Belknap Press of Harvard University Press,
2024. | Includes bibliographical references and index.
Identifiers: LCCN 2024001367 | ISBN 9780674294233 (cloth)
Subjects: LCSH: Cartography—Mathematics. | Maps. | Mathematical geography. |
Map reading. | Cartography—Social aspects.
Classification: LCC GA108.7 .R69 2024 | DDC 526—dc23/eng/20240312
LC record available at https://lccn.loc.gov/2024001367

For my grandparents—the biggest fans of my writing.

Mówiłam, że jak dorosnę, zostanę pisarką!

CONTENTS

Introduction: How to Fall in Love with Maps and Mathematics *1*
1 Curved: How to Describe the Earth *5*
2 Flat: How to Make a Map *32*
3 Scaled: How to Measure a Line *64*
4 Distanced: How to Navigate *90*
5 Connected: How to Simplify a Map *114*
6 Divided: How to Shape Society *144*
7 Found: How to Save a Life *173*
8 Deep: How to Map the Invisible *200*
Conclusion: How to Keep Up with Change *228*

Notes *233*
Further Reading *273*
Acknowledgments *279*
Illustration Credits *283*
Index *285*

MAPMATICS

INTRODUCTION

How to Fall in Love with Maps and Mathematics

I was three, maybe four years old when my parents switched off all the lights in our small but cozy flat. At once intrigued and nervous, I watched Dad turn on a small desk lamp and point it toward a cheap plastic globe, focusing the light on the East Coast of the United States as best as he could. "Look," he said, "it's dark here in Warsaw, but your aunt in New York must be just getting ready for lunch." He explained that the Earth is round and that it never stops rotating, unlike the spinning tops I liked playing with. And that, always, somewhere it is day and somewhere it is night.

After this special evening, the globe became my favorite toy. I kept spinning it, pointing my finger to places I wanted to visit, captivated by their names—from Ashgabat to Zanzibar. As a preteen, I expressed this growing passion for geography by plastering my bedroom walls with maps, saving just a little spot for a picture of a celebrity crush. It was much later that I realized that the world map on my wall and the globe, which supposedly represented the same planet, told two different stories.

On the flat map, Greenland was as big as the whole of Africa, but on the globe, the white island was dominated in size by the continent. I knew deep down that something wasn't right, but only at university, during a differential geometry* lecture, did I learn about the reason for this jarring

* Differential geometry is a branch of mathematics that studies curves, surfaces, and other shapes.

discrepancy. Even a task as simple as comparing areas of countries requires some knowledge about the mathematical principles behind the map we are using.

Flemish cartographer Gerardus Mercator created Mercator's map with the characteristic right-angled latitude–longitude grid in the sixteenth century. Though still used today, it is widely known to be deceptive to the eye. The map's area distortion conveniently supports the North-centric view of itself as being bigger and thereby more powerful. Almost half a millennium later, this map remains the standard view of the world. From primary school onward, we are fed this worldview and its intrinsic implications of the region's superiority, which impacts the way we see our home countries. Maps don't just shape our sense of space; they sculpt our perception of other nations too.

Mercator's map gets bad press these days, yet we keep an online version of it in our pockets. Its property of preserving the angles between lines on the Earth lets us easily identify the north, which makes the map as useful

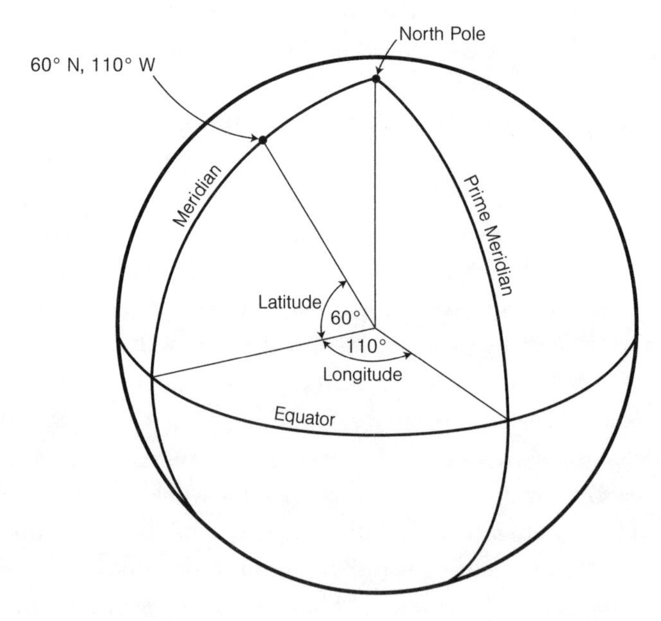

Figure 0.1 Meridians, or lines of longitude, are imaginary lines on the Earth's surface connecting the poles. Longitude specifies a point's position with respect to the prime meridian. Parallels, or circles of latitude, are imaginary circles on the Earth's surface parallel to the equator. Latitude specifies a point's position with respect to the equator.

for navigation today as it was at its birth. It should, however, come with a warning that it distorts the areas: a cartographer's equivalent of "Objects in the mirror are closer than they appear."

Mercator's map isn't distorted because of its creator's malevolence or ineptitude. In 1827, Carl Friedrich Gauss, a polymath as grumpy and eccentric as he was ingenious, mathematically proved the impossibility of flawlessly translating a three-dimensional globe into a two-dimensional map. His Remarkable Theorem (that's not a joke, but a translation of the theorem's Latin name), while jargony and full of technical assumptions, boils down to this simple fact: perfectly reducing three dimensions to two is impossible. We can't make a perfect map on a flat surface.

This book explores Gauss's Remarkable Theorem and other mathematical developments that reveal the way we make maps and, consequently, see the world. Maps represent reality, but we can take full advantage of these visual aids only when we understand the underlying math. Otherwise, we're prone to draw wrong conclusions and inherit the mapmaker's biases, whether intentional or not. For example, in Chapter 2 we'll see different distortions resulting from different ways of depicting the globe on a flat sheet of paper, and we'll gain mathematical tools to protect ourselves from being fooled by those deformations.

Beyond interpreting a map, we can use math to turn a map into a solution to a real-life problem. In Chapter 7 we'll tackle situations in which maps and math together can protect people from danger, be it a disease or a serial killer. With mathematical tools and insights, we can get much more information from a map than with our eyes and intuition alone. These math-supported applications of maps become increasingly important with improving technology and computational power.

Throughout the book, we'll see many examples of how mathematics and cartography inform each other, a relationship which has inspired the book's title. While different on the surface, the jobs of a mathematician and a cartographer are surprisingly similar. To create useful models of real-world phenomena, both mathematicians and mapmakers must choose the information to keep and omit, and different choices will lead to different conclusions. This is why we need to understand not only what we see, but what we don't see when presented with a map or a mathematical model. As we'll learn, failure to do so may lead to anything from a commuter walking a few steps too many to an international conflict.

We'll look at maps covering all scales and topics, from world maps to plans of our local streets, from counterintuitive mosque orientations to misleading underground maps. We'll visit Ancient Greece to estimate the radius of the Earth with unbelievable precision, without satellites or photography. We'll pop into eighteenth-century Königsberg, a Prussian city whose seven bridges inspired a new field of mathematics called graph theory. We'll also step into the world of fractal dimensions and, while we marvel at the surprisingly complex nature of a mundane cauliflower, we'll understand why measuring a country's borders is all but impossible and what geopolitical consequences this has.

We can't function without maps. We depend on them when we commute, travel, and interpret the news, but also when we fight diseases, catch criminals, and search for missing planes. Maps have been fueled by mathematics and have inspired numerous mathematical breakthroughs. Once we notice this connection between mathematics and cartography, we won't be able to unsee it, and it will help us to understand how our world works.

1

CURVED

How to Describe the Earth

Like so many people, I grew up believing that until the brave Christopher Columbus's "discovery" of America, people were convinced that the Earth was flat. In history classes, we were taught that in 1492, Columbus, sailing westward from Palos de la Frontera in Spain, reached what he believed to be the "Indies" (East Asia). This marked the end of the Middle Ages and the beginning of the Renaissance—or, as our textbooks claimed, the end of the Dark Ages and the start of the Age of Discovery. Without the great Columbus, our geography teachers would have been showing us continents and oceans on a pancake-shaped globe.

The myth that Columbus wanted to sail to eastern Asia to prove that the Earth was spherical seems to stem from Washington Irving's 1828 fictional biography *History of the Life and Voyages of Christopher Columbus.* The truth is, educated people have known that the Earth isn't flat since ancient times. Almost two millennia before Columbus's travels, the Greek philosopher Aristotle published *On the Heavens,* in which he mentioned that during an eclipse one can observe a spherical Earth's shadow on the Moon.* This by itself doesn't exclude the possibility of a disk-shaped Earth, but it does when we combine it with the fact that the Earth's shadow remains circular even as its orientation changes. Aristotle, noticing that the stars

*Even before Aristotle, Pythagoras had envisioned the Earth as a sphere, not a disk. We don't know, however, if this was an evidence-based idea. More likely, since ancient Greeks considered a sphere a perfect shape, Pythagoras assumed that our amazing planet must be spherical.

visible in Egypt and Greece differ, concluded that since Egypt and Greece are close to each other, the Earth must be small. Columbus wrote, "Aristotle says between the end of Spain and the beginning of India is a small sea navigable in a few days." This possibility of an easy but fruitful journey must have sparked the curiosity of this experienced (and greedy) explorer.

When Columbus reached land on his westward voyage from Spain to the Indies, confident in his navigation abilities, he announced that he had achieved what others had proclaimed impossible: he had found a faster route to India, a land rich in silks and spices just waiting to be exploited and traded. But, as we all learn in primary school, instead of Asia, Columbus had reached today's Bahamas, off the coast of North America, thus "discovering" this continent for Europeans. I for one never once questioned how such a skilled navigator could make such a huge mistake. He knew that the Earth wasn't flat, and he must surely have estimated how long the voyage would take. At first, I thought that he had mistaken the Americas for Asia because he had simply miscalculated the distance between Spain and the Indies. Then I discovered that this wasn't a mathematical mistake but a matter of data fabrication. Columbus was familiar with accurate estimates of the Earth's size—he simply chose to ignore them. Crucially, these weren't new figures coming from some cutting-edge research on measuring the Earth. The estimates were almost two millennia old.

Here Comes the Sun

Born in 276 BCE in Cyrene in ancient Greece (today's Libya), Eratosthenes was a successful mathematician, geographer, poet, astronomer, and music theorist. He moved to Egypt to become the chief librarian of the Library of Alexandria, one of the most famous libraries in history. He was also the first person known to scientifically compute the Earth's circumference, with surprising precision. Although his book *On the Measure of the Earth,* which included this result, didn't survive, it was described a few centuries later (we aren't sure of the exact time) by the Greek astronomer Clomedes.

Eratosthenes thought about the Earth as a sphere. He knew that on the day of the summer solstice, the longest day of the year in the northern hemisphere, the sun shone directly on the Tropic of Cancer, lightening the

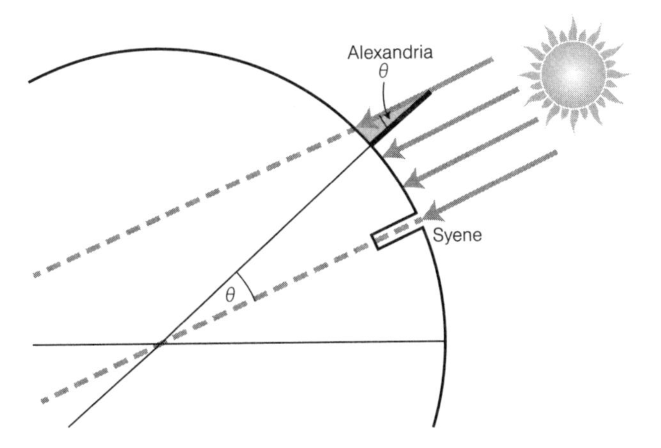

Figure 1.1 At the local noon of the summer solstice, sun rays are perpendicular to the ground in Syene and cast a shadow in Alexandria. The central angle between Syene and Alexandria and the angle between the sun rays and the gnomon in Alexandria have equal measures.

bottoms of even the deepest wells.* Knowing that at noon local time—that is, at the time of the day when the sun is at its highest point in the sky—a deep well in Syene (today's Aswan in Egypt) was lit by direct sunlight, Eratosthenes wanted to understand the position of the sun at the same time in Alexandria. By a lucky coincidence, the two cities lay along the same meridian, so the local noon occurred in both places at the same time.† To find the angle between sun rays and the Earth in Alexandria, Eratosthenes measured the angle between a vertical rod called a *gnomon* and its shadow, which turned out to be one-fiftieth of the full circle (360 degrees), so 7.2 degrees. A similar gnomon placed in Syene would cast no shadow, making this angle equal to zero degrees.

Now, imagine you're about to cut yourself a slice of pizza. Everyone knows that the best part is the crust (right?) and the bigger the central angle of the slice, the larger part of the pie's circumference—so, the more crust—you'll get. In other words, the ratio between the part of the circle's circumference between two radii and the whole circumference is equal to the ratio

*The Tropic of Cancer is a latitude of about 23°27′ N, the northernmost latitude where the sun can shine directly overhead. The Tropic of Capricorn is its equivalent in the southern hemisphere.

†Places along one meridian share the same longitude—the angle between the local meridian and the prime meridian going through Greenwich in London—and the local solar time.

between the angle between these two radii and the full circle of 360 degrees. So, the distance between Alexandria and Syene divided by the Earth's circumference is equal to the central angle between these two places divided by 360 degrees. Conveniently, the distance between Alexandria and Syene had already been measured to be 5,000 stadia, which is an ancient unit of measurement. To estimate the Earth's circumference, Eratosthenes was missing just the central angle between the two cities.

Eratosthenes assumed that sun rays are parallel to each other, which, although not technically true, for all practical purposes is a reasonable assumption. The sun is so far from the Earth and is so much bigger than our planet that only a tiny portion of the rays hit the Earth, making them almost parallel. This means that the central angle between Syene and Alexandria and the angle between the gnomon and the sun rays in Alexandria were created by a straight line (the extended Earth's radius in Alexandria) crossing two parallel lines (the extended sun rays in Alexandria and Syene). From an old theorem, still taught in geometry classes today, Eratosthenes knew that this is a pair of equal angles.* This meant that the central angle between the two cities was equal to one-fiftieth of a full circle, which allowed him to conclude that the distance between Alexandria and Syene made up one-fiftieth of the Earth's circumference. Thus, he arrived at the first scientific estimate of the Earth's circumference: $50 \times 5{,}000 = 250{,}000$ stadia.

Historians disagree on the exact definition of the stadium measurement, which makes it impossible to assess the accuracy of Eratosthenes's estimate. That exact value aside, most researchers still agree that he got astonishingly close to the actual value of about 40,000 kilometers, or 25,000 miles. He was quite lucky, having made several errors along the way that canceled each other out. For example, Syene didn't lie exactly on the Tropic of Cancer, but slightly to the north of it. Also, contrary to his assumption that Syene and Alexandria were on the same meridian, the latter lay to the west. However, mathematical models never perfectly reflect reality. What's most important is that Eratosthenes's method was scientifically sound and, had he had access to more accurate measurement tools, his estimate wouldn't have differed from the current best knowledge. That's why he's often considered the founder of scientific geodesy, which is the science of measuring the Earth's shape, orientation in space, and gravitational field.

* Precisely, this is a pair of alternate interior angles.

Lies, Damned Lies, and Columbus

We left Christopher Columbus pondering his trip to the Indies. According to the scholar Ferdinand Columbus, who happened to be Christopher's son, the explorer was familiar with the work of ancient and medieval geographers, including Eratosthenes. He used this knowledge to persuade others of his idea to get to the Indies by traveling westward. However, aware that any reasonable potential funder would consider this journey insane, Columbus carefully picked convenient facts and figures to present his argument.

To estimate the length of the potential westward journey to East Asia, Columbus needed two pieces of information: the Earth's circumference and the width of Asia. The length of his journey would be close to the difference between these two values, so the smaller the Earth's circumference and the wider Asia, the shorter the route to the Indies.

One of Columbus's greatest inspirations was Marco Polo, a Venetian merchant who at the end of the thirteenth century traveled to Asia. Columbus learned about Polo's journey from a contemporary, a Florentine mathematician and astronomer named Paolo dal Pozzo Toscanelli, who also went by the name of Paul the Physician. So respected was his knowledge of geography that Afonso V, the king of Portugal, asked him about the fastest route to the "land of spices," that is, India. Back then, spices were as precious as toilet paper in early 2020, so the king considered exploring easier ways to get to Asia a worthy investment. He wanted to know if he should send people to sail around Africa or if going westward was a better idea. Toscanelli suggested a westward route, supporting his opinion with a nautical chart he had made. Although this map didn't survive, in the following centuries multiple researchers recreated it based on later maps that had been influenced by it as well as Toscanelli's detailed description of the westward route in the original letter.

Columbus, having learned about Toscanelli's ideas for sailing west, wrote to him, asking about the details. The Florentine physician replied, attaching a copy of his letter to the king, together with the nautical chart that explained the route's details. Columbus, confident (perhaps overconfident) in his sailing abilities, wasn't looking for information but for confirmation of his beliefs from a respected scholar—and he got what he wanted. Columbus and Toscanelli both believed in the accuracy of Marco Polo's journals, in which Asia appeared thirty degrees of longitude wider than the estimates of contemporary scholars. To make things even more convenient, Polo had put the legendary rich island of Cipangu (today's Japan) over 2,000 kilometers east of Asia's coast, although its true shortest distance to the mainland is only about 200 kilometers. Despite these imaginary shortcuts, Toscanelli still estimated the journey to be about 9,000 kilometers—way more than any fifteenth-century sailor could handle. But Columbus's talent for data fudging didn't disappoint.

Before Columbus, multiple scholars had attempted to measure the Earth's circumference. As we've seen with Eratosthenes, who did a great job with the tools at his disposal, even the best methods didn't guarantee a perfect estimate, and a big issue was that different scholars used different units. Columbus decided to pick and choose his data to make the Earth as small as possible and the Indies as near as possible. In the end, he chose the estimate of the ninth-century Arab geographers who found the Earth's circumference to be equal to 20,400 Arabic miles, each mile about 2,164 meters long.* This would make 44,146 kilometers, which was close to today's value, but way too large a number for Columbus's taste. So, he took the figure of 20,400 but claimed the unit to be not the Arabic but the Roman mile, which was equal to 1,480 meters—making the Earth's circumference only about 30,192 kilometers. With these questionable calculations, Columbus reduced the Earth's circumference by about a quarter. Even on this shrunken planet, the nonstop journey between the starting point in the Canary Islands and Cipangu would be too long for the most advanced ships. So, Columbus added in some islands, all in spots perfect for breaks on the way to the destination, thus drastically reducing the longest stretch of continuous sailing. All these miscalculations brought Cipangu to approximately the same meridian as the Virgin Islands in the Caribbean, which likely impacted Columbus's notorious confusion of continents.

Finally happy with his estimates, Columbus applied to João II, the new king of Portugal, for funding for the expedition. João's royal mathematicians quickly spotted Columbus's fabrications and realized that his plan wasn't achievable. Undeterred by the rejection, he then tried his luck in Spain, where he was dismissed at least twice. But, after years of listening to Columbus's arguments, King Ferdinand II and Queen Isabella I eventually approved his proposal, mostly thanks to the enthusiasm of the Spanish treasurer.

*They reported the length of one degree of latitude but, for clarity, I have multiplied it by 360 degrees to get the Earth's circumference.

Columbus was lying to everyone else, but why was he lying to himself? In the end, he was the one who would pay a high price if the journey ended in a fiasco. As American historian Samuel Eliot Morison, author of the Pulitzer Prize–winning biography of Columbus, *Admiral of the Ocean Sea,* observed, "Columbus's mind was not logical. He knew he could make it, and the figures had to fit." He died believing—at least officially—that he had discovered the westward route to the Indies, having come across the Bahamas after thirty-three days, about the same time he had expected to reach Cipangu. According to Morison, had there been no land between Spain and the Indies, Columbus likely wouldn't have made it to Asia anyway because his ships weren't advanced enough for such a long journey. Columbus was lucky, which can't be said of the Americans decimated by European "explorers."

If a navigator as skillful as Columbus couldn't tell the Americas from Asia, how could we even dream of making accurate maps? A breakthrough in measuring techniques came a few decades later, hidden in an appendix to an initially unsuccessful book that turned into a bestseller.

The Magic of Triangles

In 1524, just a few decades after Columbus's travels, the young German mathematician and budding scientific publisher Peter Apian wrote and published *Cosmographia,* a textbook on topics ranging from astronomy to cartography to mathematical instruments. Despite its impressive scope, it wasn't a huge success. Five years later, however, a Dutch mathematician, Gemma Frisius, lightly edited Apian's book, and in this second edition, it became a popular introduction to scientific subjects such as astronomy and mathematics. *Cosmographia*'s success encouraged Frisius to publish new editions, into which he snuck some of his own work as appendixes. The appendix of the third edition contained a detailed description of triangulation, a technique that changed mapmaking forever.

In times of GPS, we take measuring long distances for granted. But in the sixteenth century it was much easier to find angles than distances, which inspired Frisius's idea.* Since measuring distances is hard, he thought, wouldn't it be nice to have to do it only once, and then calculate other distances mathematically?

* Since the publication of *Cosmographia*'s third edition, the method of triangulation has been improved. Here I'm describing the modern version, based on the same principles as the original.

In the first step, a person conducting triangulation—a surveyor—measures the distance between two known points, known as the baseline. This used to be a physically demanding task, requiring moving a long, often heavy measure in a straight line, regardless of obstacles encountered along the way. Frisius noticed that even when the goal was to survey a large area, it was enough to measure only one distance. The rest would come from trigonometry.

Trigonometry is all about studying relationships between angles and side lengths in triangles. The surveyor creates a triangle, with the baseline as one side and the third vertex at a point visible from the other two—for example, a tower or a mountain peak. Then, she measures the angles at the two observation points at either end of the baseline. Thanks to trigonometry, this is all the information needed to calculate the distances between the observation points and the point of interest, which has saved generations of surveyors from two arduous journeys through thick forests, mud, or lakes between the observation points and the third vertex of the triangle. With the power of trigonometry, the surveyor can find two distances without leaving the safe, carefully chosen baseline.

This process by itself would significantly reduce the surveyor's work, but Frisius took the idea even further. He suggested building a triangulation network, where each calculated side of a triangle would become a baseline

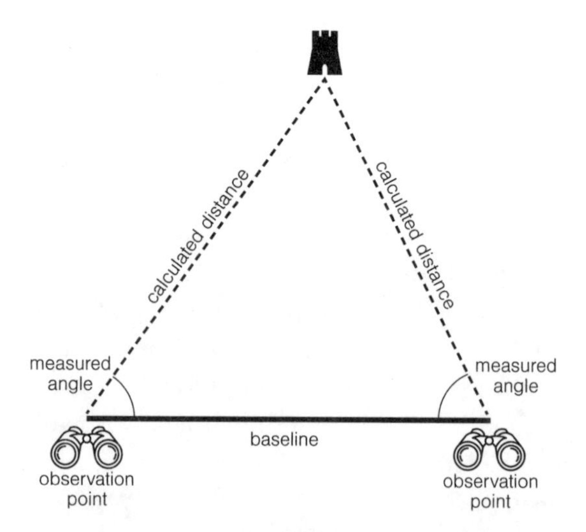

Figure 1.2 Triangulation is used to calculate the distances from two observation points to a point of interest, given the known distance between the observation points and the angles measured from the observation points.

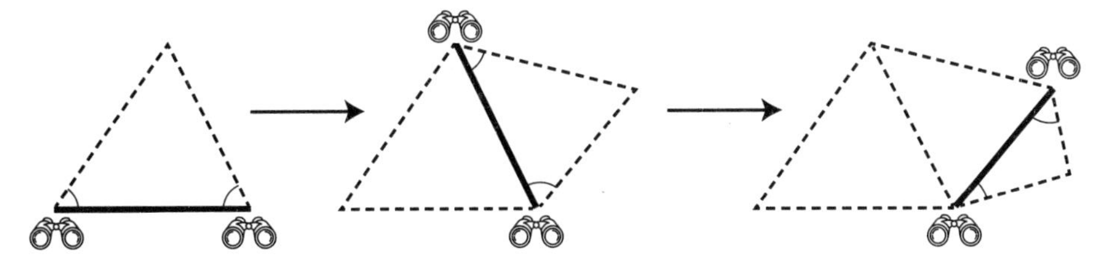

Figure 1.3 In a triangulation network, a calculated distance becomes a baseline of a new triangle.

of a new triangle. In principle, this would allow the surveyor to build an accurate map of a whole country without measuring a single distance beyond the baseline. In practice, one would measure additional distances to correct for inevitable measurement errors that would quickly add up in the process, but this wasn't strictly necessary.

Due to the popularity of *Cosmographia,* translated from Latin to French, Dutch and Spanish, the idea of triangulation quickly spread through Europe. Cartographers started using triangulation to create accurate maps, including Frisius's famous student Mercator, who was surveying the duchy of Lorraine (today's northeastern France). We'll meet him again in the next chapter.

At the beginning of the seventeenth century, another Dutch mathematician, Willebrord van Royen Snell,* took triangulation to another level by applying it to measure the size of the Earth. Although he described his project in a book aptly named *Eratosthenes Batavus* (which translates to *The Dutch Eratosthenes*), his method differed from the ancient measurement by Eratosthenes. Snell used triangulation to find the accurate distance between Alkmaar and another Dutch city, Bergen-op-Zoom, about eighty miles almost directly south. Then, using astronomical observations, he found what fraction of the Earth's circumference the stretch between Alkmaar and Bergen-op-Zoom comprised. From these two values, he calculated the Earth's circumference within 4 percent of the modern estimate, which, again, is impressive given the unsophisticated measurement and mathematical tools he had at his disposal.

* His name appears in many forms: apart from Willebrord Snell, we also have Willebrord Snel and the Latin version, Willebrordus Snellius. I'll stick to Snell, given that many readers will have learned about Snell's law, which describes the behavior of light passing between two media, such as air and water.

Triangulation, a technique at once simple and powerful, not only allowed us to make accurate maps but also helped us establish the true shape of the Earth. The Earth not being flat doesn't mean that it's spherical—but to figure that out, we needed the first-ever international expedition.

Much Ado about the Shape of the Earth

The world-famous naval architect, historian of science, and Pulitzer Prize finalist Larrie D. Ferreiro is eager to discuss geodesy, especially—he admits—with a fellow Imperial College alumna. Ferreiro's job requires a deep understanding of politics and current events, so I'm not surprised that the moment he learns I'm in Warsaw, he has many questions about the impact of the war in Ukraine on the situation in neighboring Poland. This leads us to a fascinating conversation about the importance of bringing political and social context to discussions about science, something which he has done successfully in his book *Measure of the Earth.*

By the seventeenth century, Europeans suspected that the Earth wasn't a perfect sphere, but they couldn't agree on the exact shape. The French philosopher René Descartes claimed that our planet was elongated at the poles, which would give it the shape of an egg.* On the other side of the English Channel, the British scientist Isaac Newton was arguing that the forces acting as the Earth spins flatten it at the poles and make it bulge along the equator, so the planet resembles a grapefruit. In 1687, in his groundbreaking *Philosophiæ Naturalis Principia Mathematica* (*Mathematical Principles of Natural Philosophy*), Newton discussed the theory of gravity, an almost magical attracting force. It was a difference in gravity, Newton believed, that caused pendulum clocks to beat more slowly at the equator than in Europe. He argued that the closer a clock was to the center of the Earth, the larger gravitational force was acting on the pendulum, and the faster it was beating. This difference could only be explained by a flattened planet.

Far from an obscure scientific debate, deciding whether the Earth was egg- or grapefruit-shaped had strategic importance. The nation better at navigation would build a stronger empire, so neither the British nor the

*In his 1644 *Principia Philosophiae* (*Principles of Philosophy*), Descartes proposed that the planets, the stars, and the Moon are all surrounded by an invisible fluid called "ether," swirling ever since God started the circular motion, which was supposed to explain the Earth's shape.

French could afford their ships getting off track by hundreds of miles, which might happen if navigators at sea assumed the wrong shape of the Earth. Winning a scientific argument also had symbolic meaning—each of the two superpowers wanted to prove the theory of *their* scientist: France supported the Cartesian egg shape, while Britain favored the Newtonian grapefruit.

During our conversation, Ferreiro compares this debate to the Cold War race to the Moon, which "had nothing to do with science." The United States and the USSR each believed that whoever got there first would show the world which country and, by extension, which system—capitalism or communism—was worth supporting. Similarly, the battle over the shape of the Earth was as political as it was scientific. Paraphrasing a Prussian general, Carl von Clausewitz, Ferreiro argues that "science is the continuation of politics by other means." He explains that "science by itself certainly is the search for facts, but if you step back and look at the important things—like who's paying for it, who's funding it, what is the intent of that support—you find very quickly that there [has] always, always been a political dimension of science."

Jean-Frédéric Philippe Phélypeaux, comte de Maurepas, a young but talented minister in the court of Louis XV, understood the political dimension of science. So, when in 1734 the French Academy of Sciences received a proposal that would settle the debate about the shape of the Earth once and for all, Maurepas became its greatest supporter. This ambitious project involved sending a mission all the way to the equator to measure a degree of latitude.

Both the northern and southern hemispheres are divided into ninety degrees of latitude—from zero degrees at the equator to ninety at the poles. If the Earth were a perfect sphere, the length of one degree of latitude would be the same everywhere and equal to the Earth's circumference divided by 360, so to about sixty-nine miles. Since the Earth isn't spherical, however, one degree of latitude covers a different distance, depending on where it's measured. This means that by comparing one degree of latitude at the equator and the already-measured degree of latitude in France, the French scientists would be able to figure out whether the Cartesian or Newtonian view of the world was correct. On an egg-shaped planet, a degree of latitude in France would cover a longer distance than a degree of latitude close to the equator, while on a grapefruit-shaped planet, the opposite would be true.

Once the idea was approved, Maurepas had to choose the optimal destination for the scientific mission. He quickly rejected the hostile equatorial African coast and the remote tropical Asian islands, settling on a Spanish colony in South America: Peru. The king of Spain, who happened to be the uncle of France's Louis XV, gave France his blessing to conduct their measurements on Spanish territory. To ensure their access to all the scientific knowledge gained during the expedition, and to reduce the chance of the French smuggling goods out of Peru, Spain insisted that "two intelligent Spaniards accompany the said scientists." After intense preparations, in spring 1735 a team comprising both French and Spanish members set off to South America, marking the beginning of an unprecedented international scientific expedition.

The group of academics, accustomed to carrying out theoretical research from their comfy armchairs, was unprepared for the challenges awaiting them in Peru. They didn't expect the extreme cold of the high peaks of the Andes, the understandable hostility of Peruvians toward European invaders, or the local politics. Most of all, they had estimated that the mission would take two years. Little did they know when they set out that the first member of the group wouldn't return home for over a decade, and some wouldn't return at all. Given how unfit for the task the mission participants were, it's a miracle that the mission succeeded.*

Using triangulation, the expedition aimed to find the distance between Quito in the north and Cuenca in the south. The physical demands of the job surprised the surveyors. The distance between the two cities, both in today's Ecuador, was about 200 miles, approximately the same as from London to Paris or from Boston to New York City. But this wasn't the worst part. As triangulation requires a network of easily observable points, in practice, it meant going up and down the peaks of the Andes, although the scientists didn't have it as bad as the local people forced to carry heavy equipment for their white "employers." To ensure the precision of measurements, they used an iron instrument called a *quadrant,* which was reliable but cumbersome, with a radius of up to three feet. And getting up a mountain not only was difficult but also didn't guarantee good visibility: often, the team had to wait for a clear moment for days, even weeks, in rain or snow.

*The mission initiator and formal leader, Louis Godin, was more interested in local women than measurements and ended up spending a large portion of the mission's money on a diamond for one of his lovers.

And there was lots of snow, which wasn't what the surveyors had signed up for on a trip to the tropics!

Before the real triangulation even started, the team spent months measuring the seven-mile baseline, which, although conducted on flat terrain, may have been even more arduous than climbing the peaks. They would start by placing an iron rod over six feet long, called a *toise,* at the beginning of the baseline and marking where the toise ended. They then moved this heavy object along the baseline, each time starting from the previous end point. They repeated the process thousands of times until they reached the end of the baseline. And then, as a last step of triangulation, they repeated the whole process. This second time, while measuring the baseline wasn't strictly necessary, it allowed them to assess the accuracy of their triangulation.*

When the triangulation was complete and the physical labor over, the scientists regrouped in Cuenca for a long period of complicated mathematical calculations. As we've seen, to find the side lengths of a triangle given the length of the third side and two angles, one needs to apply trigonometry—and the process had to be repeated for every triangle in a chain over 200 miles long. To ensure greater accuracy, they applied a variety of corrections to compensate, for example, for the differences in altitudes of observation points as well as the Earth's curvature,† which further complicated the calculations.

Finally, astronomical observations allowed them to figure out the latitudes of Quito and Cuenca. After over two years of measuring the position of stars in the sky and sophisticated mathematical calculations, they arrived at the angular distance between the southern and the northern end of the triangulation chain. By dividing the result of the triangulation by this value, they calculated the exact length of one degree of latitude at the equator to be 362,899 feet, which is within 120 yards of the currently accepted value. Shorter than one degree of latitude in Paris, this result proved the Newtonian theory of the Earth flattened at the poles.

The implications of the Geodesic Mission to the Equator went beyond establishing the true size of the Earth. The mission's success showed future generations of scientists that international collaboration was possible, and

*To make things even more complicated, they measured two baselines after the triangulation because Godin and the rest of the team couldn't agree which one to measure.

†This was the first time that the Earth's curvature was accounted for in triangulation.

so were ambitious research projects in mostly unexplored lands, which inspired the groundbreaking expeditions of polymaths such as Alexander von Humboldt and Charles Darwin. Moreover, decades spent in the region taught Europeans about the rich local cultures, uninfluenced by the Spanish empire, inspiring the idea of independent South American nations. Indeed, the Venezuelan political leader Simón Bolivar described this mission as an inspiration for his liberation movement. Possibly nothing speaks more to the importance of the Geodesic Mission to the Equator than the etymology of Ecuador, the name of a country that in 1830 gained independence from the Spanish colonizer: República del Ecuador means simply "Republic of the Equator."

Understanding the shape of the Earth has proved crucial in the development of mapmaking, from paper maps to today's GPS. Knowing the precise measurements of our planet, however, didn't automatically lead to perfectly accurate maps—not in the eighteenth century, and not at the time of writing. We can develop our geodetic knowledge all we want, but we will never make a flawless flat map, and the curved shape of the Earth itself is to blame.

Prince of Geodesy

The question of mapping the almost spheroidal surface of the Earth onto a flat plane was one of many research interests of Karl Friedrich Gauss. Born in 1777 in Brunswick (today's Germany) to poor parents without much formal education,* he quickly revealed his exceptional talents. At seven, he entered the local elementary school, where the principal, J. G. Büttner, motivated about two hundred students with the liberal use of a whip.

Keen to occupy the misbehaving children, Büttner reportedly instructed them to add numbers from one to one hundred.† After less than a minute, nine-year-old Gauss handed his teacher a tablet with the correct answer. Instead of adding the numbers one by one, he had found fifty pairs with a sum of 101 each: $1+100$, $2+99$, $3+98$, etc., which gave $50\times 101=5{,}050$ in

*His mother, unable to write, didn't record the date of birth of her only son. She remembered that he was born on a Wednesday, eight days before Ascension, which later allowed Gauss to apply his newly developed formula for the date of Easter to calculate his birth date: April 30.

†A former editor and columnist for *American Scientist,* Brian Hayes, tried to track down the source of this anecdote. It seems that it actually happened, but it's unclear what sum the students were asked to compute.

total. The teacher, recognizing his talent, encouraged Gauss's father to let him study in the evenings instead of helping around the house. Then, understanding that he had taught the gifted child all he knew himself, Büttner ordered more advanced arithmetic textbooks, kickstarting the extraordinary career of the future "Prince of Mathematics." But even this generous epithet doesn't convey the range of Gauss's achievements, not only in arithmetic, geometry, probability, and algebra but also in magnetism, astronomy, and cartography, to name just a few. He saw the value in applying mathematics to solve real-world problems and expressed the desire to become "the most refined geometer and the perfect astronomer." It didn't hurt that applied work, as opposed to theoretical studies, usually came with more generous funding.

Gauss realized that to make accurate astronomical observations, he had to know the precise position of the observatory and the correct shape of the Earth, which likely sparked his early interest in geodesy. A perfectionist in everything he dabbled in, he soon became a recognized expert in this field. A renowned historian of cartography, Matthew Edney, tells me that he considers Gauss "the geodetic god," adding that he laid the foundations for modern geodesy. It's not a coincidence that, despite his many achievements, it's the diagram of his triangulation network together with a sextant—the navigational instrument he used—that made it onto the 1993 German ten-mark banknote celebrating his life and work.

After years of participating in various triangulations, in 1818 he was put in charge of a geodetic survey of the kingdom of Hanover in today's northern Germany. He took his job seriously and performed many of the observations himself, as enthusiastic about gathering valuable data as he was unprepared for the fieldwork. Gauss would ride a horse in elegant but impractical clothing, including a velvet cap, which once led to such overheating that his ill health necessitated a break in the project. On another occasion, he was thrown to the ground by a horse, though he didn't suffer any injuries beyond some cuts and bruises. But Gauss was inspired rather than discouraged by such obstacles. Indeed, fellow German mathematician Friedrich Wilhelm Bessel tried to deter Gauss from wasting time and energy on the physical work of geodetic surveys. Aware of Gauss's extraordinary mathematical talent, he worried that his hands-on approach would stop him from producing theoretical results. However, his worries weren't warranted as, rather than stop him from pondering over theoretical questions, practical geodetic work inspired some of Gauss's best mathematical ideas. In particular, he wanted to understand the geometry of projecting one surface onto another, for example, a surface of a sphere onto a flat plane—in other words, mapmaking.

In 1827, Gauss presented to the Royal Society of Sciences in Göttingen his *Disquisitiones generals circa superficies curvas* (*General Investigations of Curved Surfaces*), which included the results of his geometric research influenced by geodetic surveys. Among other results, he rigorously proved the impossibility of making a perfect map, the unintended consequence of which is the proper way of . . . eating pizza (although it's unlikely that Gauss ever had a chance to try this delicious dish).

Pizzas and Bananas

A picky Italian, my date felt the need to drag me across half of London to the one and only acceptable pizzeria. As soon as I picked up a slice of pizza, a delicious layer of tomato sauce and olives fell onto my lovely yellow shirt, which I had bought especially for the occasion. My face turned the color of said tomato sauce, and I was sure that our first date would also be the last. My date was a mathematician, so why, I wondered, would he want to go out with someone who dared to eat pizza in such a nonmathematical way? What caused this literal flop wasn't my clumsiness, however, but my ignorance of Gauss's powerful theorem about curved surfaces—the same one that explains why all maps are distorted.

One of the simplest surfaces we deal with in everyday life is a flat sheet of paper. If we roll it into a cylinder—like a newly purchased poster, ready to be transported—this previously flat page remains flat in the vertical direction but curves in the horizontal direction. So, is a sheet of paper flat or curved?

To reason about surfaces, mathematicians use curvature, which describes the behavior of a surface at a specific point. After picking a point on a surface, we draw a line through it and quantify how much it curves. Let's consider a banana, as in Figure 1.4, and look at a point at the top of its surface. The line along the banana's "smile" curves inward, making the curvature negative. In the perpendicular direction, the line curves outward, making the curvature positive.* A good example of a zero curvature is a

*The assignment of positive or negative curvature depends on our frame of reference, so we could say that a line curving inward has positive curvature, while a line curving outward has a negative curvature. What matters is whether the two lines have curvatures of the same or opposite signs.

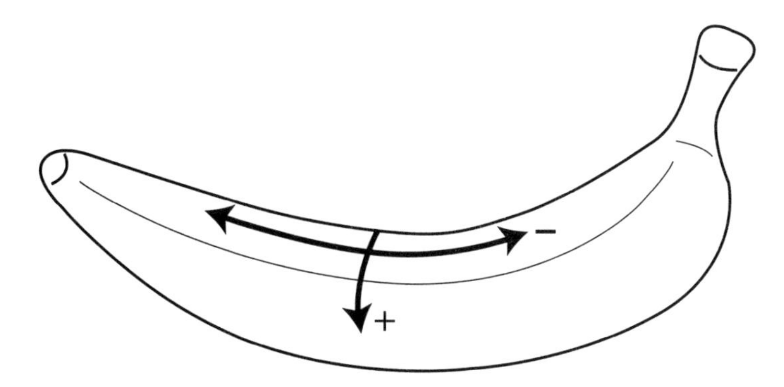

Figure 1.4 A line that curves outward has a positive curvature, while a line that curves inward has a negative curvature.

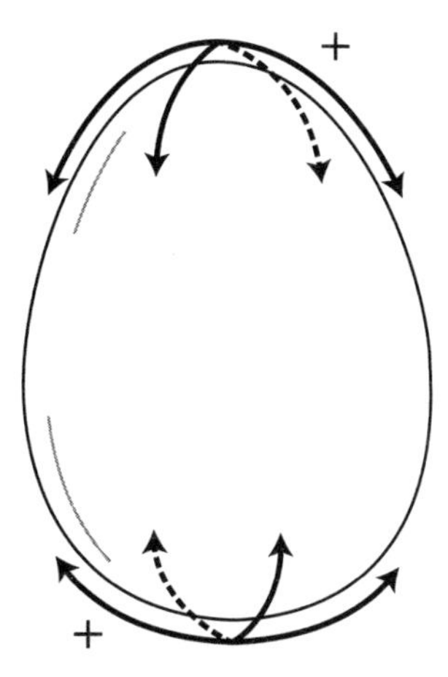

Figure 1.5 Both at the top and on the bottom of an egg, all lines have positive curvatures, but the curvatures at the pointier top are larger.

flat sheet of paper, where all lines are flat. While the sign of curvature indicates the line's general behavior, its magnitude tells us how much it's curved. For example, while both at the top and on the bottom of an egg all lines curve outward, the lines passing through the pointier top have larger curvature, as seen in Figure 1.5.

If we pick a point on a cylinder, things get more complicated. Figure 1.6 shows that the horizontal line going around the cylinder curves outward, which makes its curvature positive, while the vertical line is flat, so it has a curvature of zero. All other lines form helices with positive curvatures—the more horizontal, the higher the curvature's magnitude. So what is the

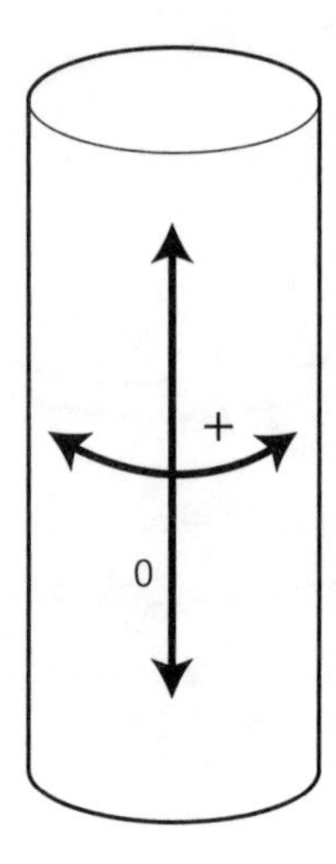

Figure 1.6 On a cylinder, horizontal lines curve outward, and vertical lines are straight.

curvature of the point on a cylinder—zero or positive, and if the latter, how big? This is a question that Gauss wanted to answer.

The Remarkable Theorem

Gauss figured that because it doesn't make sense for the same part of a surface to be convex, concave, and flat at the same time, we shouldn't be able to assign different curvatures to the same point. He devised a procedure to express the curvature of a point with a single number. Each possible line going through the point on our surface has curvature assigned to it, and we can choose the smallest and the largest of these numbers. To obtain the Gaussian curvature, we multiply the two numbers. This way, we reduce the curvatures of all the different lines passing through the point to a single value. Be careful not to confuse the curvatures of lines with the Gaussian curvature of a surface!

On the top of an egg, for example, all lines have the same, positive curvature. Their product is positive too, so the top of an egg—or any point on the egg, for that matter—has positive Gaussian curvature. At the point on the top of our banana, the largest curvature is positive, but the smallest curvature is negative, which gives a negative product. On the flat sheet of paper, all lines have zero curvature, and zero times zero gives a Gaussian curvature of zero. On the cylinder, the largest curvature—of the horizontal line—is positive, and the smallest—of the vertical line—is zero, which makes their product zero. And that makes sense, as the cylinder is formed by bending a flat sheet of paper.

Gauss understood that the Gaussian curvatures of a flat sheet of paper and the same sheet rolled into a cylinder weren't equal by coincidence. He showed that the Gaussian curvature doesn't change even if we bend the surface, as long as we don't stretch, shrink, rip, or destroy it in any other way. The nondestruction requirement is essential, and soon we'll discover its real-world consequences. Gauss was so proud of his result that he named it *Theorema Egregium*, which is Latin for "Remarkable Theorem." Indeed, this theorem has deep consequences not only in theoretical mathematics but also in activities as mundane as eating pizza.

Traditional pizza is so thin that it resembles a two-dimensional surface. Because a slice of pizza is easier to bend than stretch, shrink, or rip, we can look at it through the lens of the Remarkable Theorem. As Figure 1.7a shows, when a slice of pizza lies flat on the plate, at all points curvatures in all directions are zero, which makes their product—the Gaussian curvature—zero. This corresponds to the flat sheet of paper. When I made the mistake of picking up my slice by the crust, gravity forced the tip of my slice—together with the toppings—to bend down, as shown in Figure 1.7b. Although this changed the curvature of points in the direction from the tip to the crust, it didn't violate Gauss's Remarkable Theorem since the perpendicular line stayed flat, keeping the product zero. Instead, I should have bent my pizza in half, as in Figure 1.7c. Then, in one direction the curvature would have become negative, but since the Gaussian curvature of the slice must remain zero, the slice would have stayed flat in the direction pointing toward my mouth. This is because the only number that gives zero when multiplied by something negative is zero.

A slice of pizza that curves in one direction becomes stiff in the other direction, to keep Gaussian curvature at zero. As I was bending my pizza slice (and my mind) to understand this idea, I realized that it has surrounded us from time immemorial. Take a tree leaf, for example, which tends to fold along the central vein, like a properly handled slice of pizza. By creating a nonzero curvature in the vertical direction, it stiffens in the horizontal direction, making it harder for the wind to change its shape. This also explains why leaves that grow faster on the sides than at the center must wrinkle along the edges to keep the folded shape.

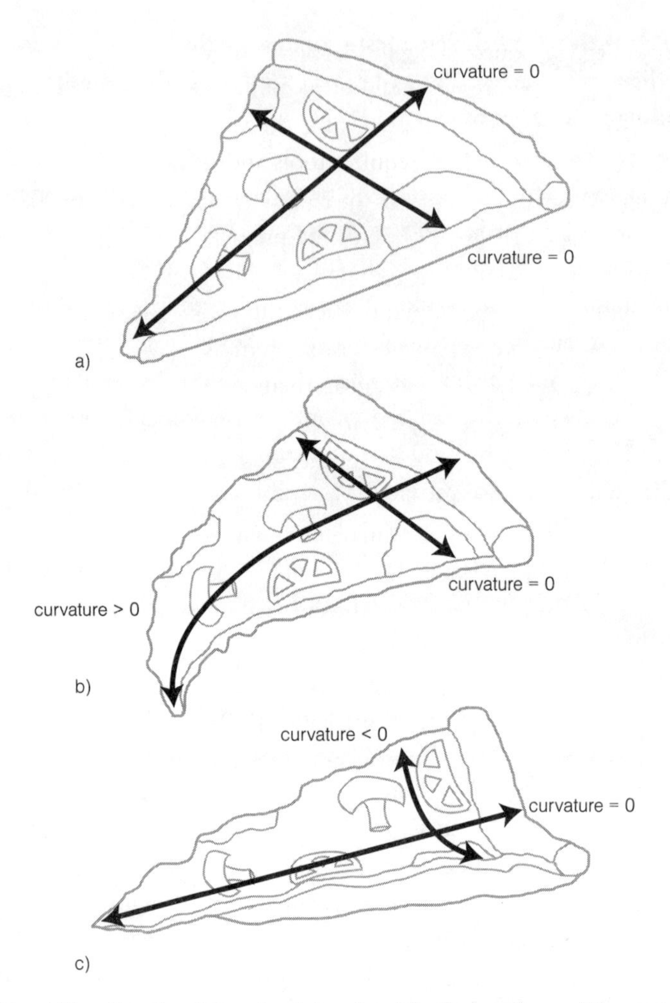

Figure 1.7 *a)* When the slice of pizza lies flat on the plate, the curvatures of the marked point are zero in all directions; *b)* If we pick up the slice of pizza by the crust, the smallest curvature stays zero, but the largest curvature becomes positive, forcing the slice to bend down; *c)* If we fold the slice of pizza, the smallest curvature becomes negative, but the largest curvature must stay zero to keep the Gaussian curvature zero. This prevents the slice from bending down.

Changing Gaussian curvature requires force, which is what makes curved objects so strong. While a folded slice of pizza and a blade of grass are curved in one direction and protect the zero curvature in the other direction, objects curved in all directions are almost unbreakable. That's one of the explanations behind the strength of the curvy egg shape. Eggs might seem fragile and, indeed, dropping one on the floor is a guaranteed mess. But try breaking an egg by squeezing it and you'll see how its curvature protects the fragile shell. Otherwise, eggs would break under the weight of a bird sitting on them, which would be disastrous for the species. To break an egg, you need to dent the shell, which requires tools and intent (so make sure to remove any rings when you squeeze the egg to test the hypothesis).

The Remarkable Theorem explains the prevalence of curved surfaces not only in nature but also in engineering and architecture. My favorite example is Pringles, the popular potato chip. Packed in tubes, flat chips would break under the weight of the chips on top of them. The characteristic shape of Pringles, curved in two directions, gives them exceptional strength. The lack of a weak point makes them easy to store and ensures that they break at random points when we bite them, increasing the sensation of crunchiness. If you're still doubting the strength of Pringle-shaped objects, look at the saddle roofs of London's VeloPark, Scandinavium in Gothenburg, or L'Oceanogràfic in Valencia. Thin, safe, and beautiful—that's the power of curved structures.

Alas, the difficulty of changing the curvature of a surface also has negative consequences. To make a map, we translate a portion of an almost spherical surface of the Earth with positive curvature onto a flat sheet of paper with curvature zero. The Remarkable Theorem tells us that this is impossible without stretching or tearing the surface. That's why sticking a Band-Aid on a knee or an elbow is so annoying, and why wrappers of spherical lollipops have wrinkles. Of course, this hasn't prevented us from creating maps, but every single flat map is distorted in some way—either distances, shapes, or areas are off. In the next chapter, we'll investigate this problem in detail.

Fasten Your Seat Belt

The curved shape of the Earth not only prevents us from making perfect flat maps but also renders a big part of school geometry invalid. The facts about angles, lines, and triangles that generations of students have been tested on quietly assume a flat surface. Geometry gets more complicated the moment the curvature of the surface changes from zero.

A popular brainteaser introducing issues around the Earth's curvature tells the story of a hunter who leaves her tent, walks ten miles south, then ten miles west, then ten miles north, where she spots a bear standing next to her tent. What color is the bear? the puzzle asks. I encourage you to think about it for a minute before reading the solution in the next paragraph.

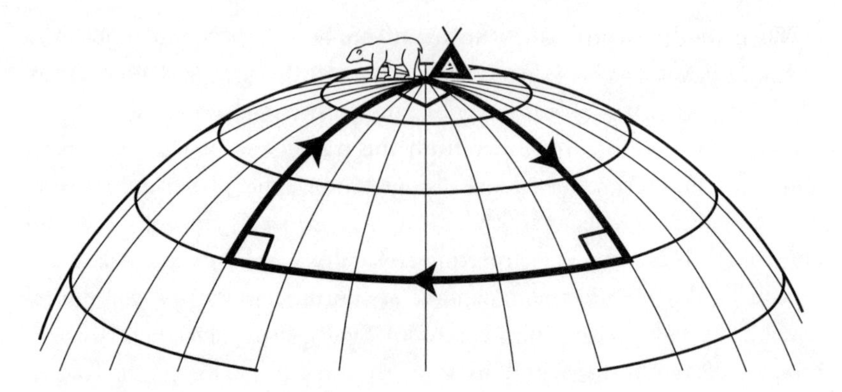

Figure 1.8 If the hunter starts at the North Pole and walks ten miles south, ten miles west, and ten miles north, she'll arrive back at her tent.

The hunter's tent must stand at the North Pole, meaning she sees a white polar bear.* If she starts at the North Pole, then walks ten miles south, turns clockwise by ninety degrees and walks ten miles west, again turns clockwise by ninety degrees and walks ten miles north, she'll return to her tent, as shown in Figure 1.8. This seemingly irrelevant puzzle about the color of a bear's fur illustrates the weird geometry of a sphere.† The hunter's path is a triangle with two angles equal to 90 degrees. When we add the third angle, the total will be larger than 180 degrees, going against the math teachers' adage that angles in all triangles add up to 180 degrees. This rule works on a plane, but spherical triangles are more interesting creatures.

On a sphere, the larger the triangle, the larger the sum of its angles—always larger than 180 degrees. This is another explanation, beyond the Remarkable Theorem, of why we cannot make a perfect map. It would require translating spherical triangles to plane triangles, which is impossible without distortions since their angles add up to different values.

* The North Pole isn't the only point on the Earth where this could take place, but it's the only one where the hunter is likely to see a bear. All other solutions place the hunter close to the South Pole. For example, she could have started anywhere on the parallel ten miles north of the ten-mile-long parallel close to the South Pole. You might enjoy trying to find other possibilities—a picture will help!

† To make things simpler, we assume here that the Earth is a perfect sphere; its true shape has similar geometric properties.

When we discuss triangulation we encounter trigonometry, which describes relationships between angles and lengths in triangles. After seeing the weird angles of spherical triangles, you won't be surprised that the rules of spherical trigonometry differ from the traditional, flat trigonometry taught at school. This means that distances on a sphere behave differently from distances on a plane.

On a plane, the shortest distance between two points is a segment of a straight line, but all lines on a sphere are curved. So, "straight" lines on a sphere are *great circles,* that is, arcs of imaginary circles whose radius is the sphere's radius. In other words, these are the largest possible circles that can be drawn on the surface of a sphere, such as an equator. Keep in mind that on a plane we can draw only one straight line between two points, while on a sphere, every two points are connected by two arcs of the same circle—the shortest distance will be the length of the shorter arc.

This fact often confuses passengers on long-haul flights. On a recent flight from Munich to San Francisco, my seat neighbor was surprised if not outright shocked when we had the opportunity to look down on the picturesque snowy mountaintops of Greenland. He stopped his movie and started poking the in-flight entertainment screen to find the flight tracker, which showed that we were following a rainbow-shaped path over Greenland instead of the expected straight line over the Atlantic.

My neighbor's confusion was understandable. San Francisco has a similar latitude to Seville in southern Spain. Munich, while to the north of both, is still considerably south of snowy Greenland. While I don't know much about aviation, I doubt the airline company is generous enough to invest in fuel to offer its customers a sightseeing tour. So why did we make this weird circle above the Earth, instead of going as the crow, or rather an airplane, flies?

Because all flat maps are distorted, what looks like a straight line on a map isn't the shortest route. The best way to see that is with a globe and a piece of string. Attach one end of a piece of string to Munich and, keeping the string as tight as possible, find San Francisco. By keeping the string tight, you ensure that the route is the shortest possible—and you'll see that it's exactly like the rainbow in Figure 1.9.

Airlines schedule their flights as close to great circles as possible because these are the shortest routes joining two points on the Earth. While actual routes will deviate from great circles to take advantage of strong winds, avoid difficult weather conditions, and stay out of no-fly zones, they still don't

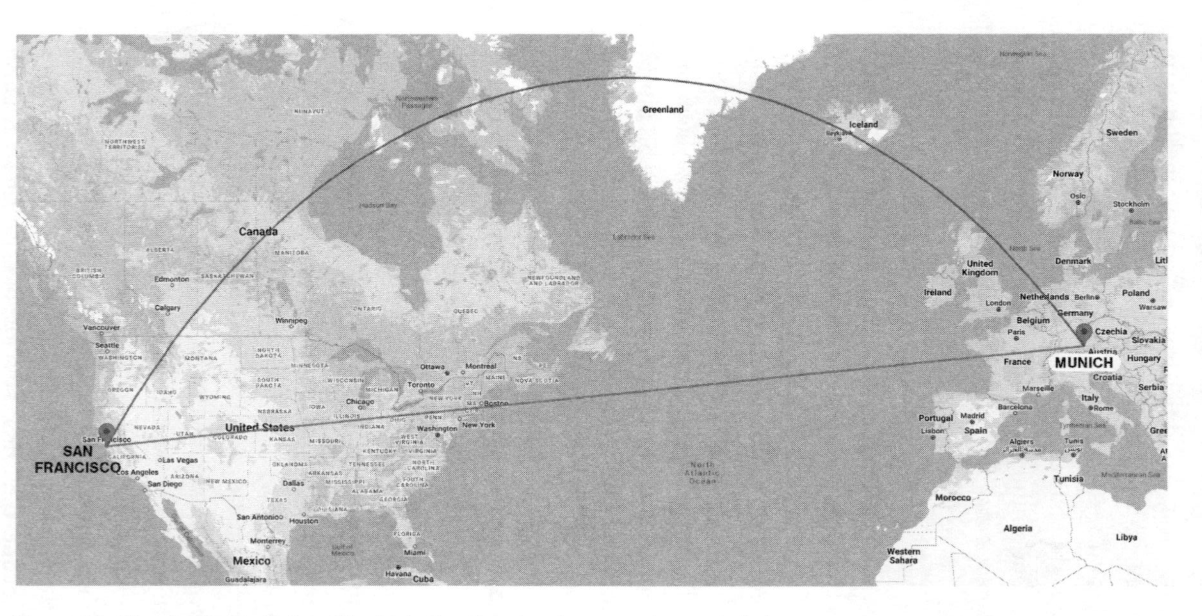

Figure 1.9 The shortest path from Munich to San Francisco is an arc of a great circle.

resemble the straight lines we're so tempted to draw on a flat map. This explains why airlines sometimes place their hubs in rather surprising places.

Flying through Alaska

For one day, on April 25, 2020, the international airport in Anchorage, Alaska, became the busiest in the world. It's surprising for a city with a population of around only 300,000, and with just over 700,000 people living in the whole, vast state. When it comes to airports, however, it's great circles that matter.

Anchorage benefited from the rapid development of aviation in the twentieth century, the increasing economic power of Asia, and maybe unexpectedly, the Cold War. Bearing in mind great circles, the shortest flight between Europe, let's say London, and East Asia, let's say Tokyo, would go over Siberia. During the Cold War, however, the Soviet Union closed its airspace—almost the size of North America—for airlines from the West. All flights originating in Europe had to take the most efficient route avoiding the no-fly zone, and this took them over Greenland and Alaska. Back then, no plane would have been able to fly for so long without refueling. Conveniently, Anchorage happened to lie about halfway along this route, and it was the only city along the way. In 1951, an international airport was built in this small, remote city. It connected major destinations all over the world, from London, Paris, and Amsterdam in Europe to Tokyo and Mumbai in Asia, and New York and São Paulo in the Americas. By the 1980s, this airport was nicknamed the "Crossroads of the World."

After the fall of the Berlin Wall in November 1989, the USSR opened its massive airspace for most airlines. This, together with the development of modern, longer-range airplanes, diminished the role of Anchorage Airport for commercial flights. The rapid decrease in commercial flights notwithstanding, the importance of Alaska in international aviation has only grown since its heyday. While today's commercial airplanes can fly nonstop for more than the 9,500 miles between Singapore and New York, long distances pose a challenge for cargo flights. For them, it's a trade-off between taking more fuel, increasing the range, or taking more cargo, increasing the revenue per flight. To maximize the weight of cargo onboard, transport companies had to find a convenient refueling stop as close to the route as possible, and Anchorage is the only airport in the world that lies close to great circles connecting dozens of major cities. Today, Anchorage is home to the hubs of giants such as FedEx, United Parcel Service (UPS), and the US Postal Service, where airplanes stop to refuel and sort the cargo, sending it on to correct destinations.* All that explains why, in April 2020, when commercial flights stopped almost entirely due to COVID-19 restrictions and the shipping of goods bought online spiked, Anchorage became one of the world's busiest airports—and, for one day, *the* busiest.

Most major airlines—both commercial and cargo—work in a hub-and-spoke system, which means that their flights start or end in one of the hub airports, where passengers change to the connecting flight to their destination. For example, at the time of writing, the world's busiest international airport is in Dubai. The biggest city in the United Arab Emirates is conveniently located along the great circles connecting London and Perth in Australia. Similarly, passengers flying from London to Mumbai in India might want to change at the large airport in Istanbul. Finally, the location along great circles between major European destinations and China has led to the success of the airport in the relatively small city of Helsinki, the capital of Finland.

Size and Shape Matter

When, in 2016, an American rapper named Bobby Ray Simmons Jr., known as B.o.B, posted a series of over fifty tweets, some including photos, to prove that the Earth was flat, the astrophysicist Neil deGrasse Tyson was having none of it. In a few tweets, he pointed out errors in the musician's line of thinking. Within hours, the discussion left Twitter, and B.o.B released a diss track, "Flatline," about his theory, with the lyrics suggesting that Tyson is paid to indoctrinate people. The scientist gave as good as he got, responding with his rapping nephew's song "Flat to Fact."

Despite centuries of incontrovertible scientific research, some people are still compelled to engage in heated disputes about the shape and size of the Earth, which shows that this topic is as important today as it ever was. Get your calculations right, and your nation stands to gain political advantage; get it wrong, and you'll perhaps end up on the wrong continent. Geodesy—and the mathematics behind it—allows us to find out where

*In Chapter 5, we'll discover what happens to these packages when they get closer to the recipient.

we are, how to get where we want to be, and how long the journey will take. But when we read flat maps of our curved planet, it's important to remember the Remarkable Theorem. The inevitable distortions created by the Earth's curvature impact our interpretation of the map, which makes understanding the mathematics of mapmaking crucial to forming a worldview that is as unbiased as possible.

2

FLAT

How to Make a Map

In March 2017, Boston Public Schools changed their pupils' view of the world. Literally. The familiar world maps on the walls of the classrooms were replaced with new ones, on which everything looked different. Overnight, students saw Europe and North America shrink, giving place to a much larger Africa and South America than they had been used to seeing. The continents looked a bit funny, stretched horizontally close to the poles and vertically near the equator. Africa became an elongated giant landmass, while Europe—usually presented as not much smaller than its southern neighbor—almost disappeared from the map. All these changes can be explained with math.

Lightbulb Moment

Cartographers face a mathematically impossible task. From Gauss's Remarkable Theorem, we know that transferring our three-dimensional globe onto a two-dimensional map is doomed to failure. Despite that, we still haven't given up on using maps, because even a distorted two-dimensional map tends to be of more practical use than a globe. Imagine a pilot in a cockpit drawing a flight route on a globe, or a hiker who has to leave her sleeping bag at home because the globe takes up too much space in a backpack, or a mini globe rolling around your car after particularly harsh braking. We're not ready to give up on maps, and we've learned to accept that they always distort some of the globe's features.

We can translate the same area of the world onto a flat sheet of paper in many different ways, known as projections. Cartographers spend a lot of time deciding what kind of projection would be most appropriate for a given application—in other words, what distortions are least relevant and what features must stay. It is our responsibility, as map users, to understand what attributes the map depicts truthfully and which it deforms, so that we don't build false images in our minds. In general, we care about preserving three main characteristics: areas, shapes, and distances. Which one will it hurt the most to lose? This is the first question we should ask ourselves before selecting the most appropriate map projection.

If preserving the area is what we want, then we should choose one of the equal-area projections (if you want a fancier name, they're also called *homolographic projections*). These have an appealing property whereby if you put your thumb anywhere on a map and then move it to another place, in both instances it will cover the same real-world area. But, as we know from Gauss, shapes and distances will be distorted in the process, in some parts of the world more than others, depending on how exactly we construct the map. The good news is that if we choose a location of the most interest to us, we can design the map so that it approximately preserves shapes and distances around this place.

Sometimes we might care more about the shape than the area, so we'll accept a tiny Africa and a huge Europe as long as Africa looks like Africa and Europe looks like Europe—which wasn't the case on the new map presented to Boston students. We call such projections *conformal* or *orthomorphic*. How can we recognize a conformal map? If the meridians intersect the equator and other parallels at right angles, like on the globe, then there's a good chance (although no certainty) that the map preserves the shapes. Unfortunately, you can't have your cake and eat it too—math prevents us from creating an equal-area *and* conformal map at the same time.

If distances are the crucial feature, we could consider an equidistant projection (no fancy Greek words here, as far as I know). Don't get too excited, though—no projection keeps the same scale on the whole map (and we'll talk more about map scales in the next chapter). Again, the Remarkable Theorem is to blame. The good news is that we can preserve distances between one (or even two) special points and any other point on the map. For example, if we center our equidistant map on London, then distances between London and Paris as well as between London and Glasgow will be preserved, but I wouldn't use it to plan a trip between Paris and Glasgow. A common choice is to keep constant the distances from one of the poles along all meridians. In this case, the other pole is stretched to a circle forming the map's edge, as on the flag of the United Nations.

In practice, how do we go about flattening the Earth? Imagine you place a lightbulb in the center of a transparent globe with the latitude–longitude grid and country contours on its surface. If you wrap a sheet of paper around the globe, the light will project the grid and country contours onto it, so you can trace them with a pencil and unwrap the paper to get your map projection.

The question is, how should you wrap the paper around this globe? Cartographers face this dilemma when choosing which projection to use, and different choices will result in completely different presentations. We divide map projections into three main families, depending on how we place the paper around our transparent globe. One way is to fully wrap the sheet of paper around it, creating a cylinder touching the globe along a circle, as in Figure 2.1a. We call the resulting projection cylindrical. Because you can choose the circle along which the paper touches the globe in any way you wish, the cylindrical family contains infinitely many map projections. As you can imagine, the most popular choice of a circle is the equator, but this doesn't mean that other circles create worse projections—just different. All resulting maps are rectangular, and the right edge of the map is continued on its left.

Let's say we decide to go mainstream and choose the equator as the circle that touches the paper. What happens to the parallels and the meridians? Well, the equator is already touching the paper, so it must become a straight line on the map. The other parallels will appear as straight lines parallel to the equator, while meridians will appear as straight lines perpendicular to it. In particular, the projection will preserve all the angles in the latitude–longitude grid—right angles on the globe.

Another option is to roll the paper into a cone and place it like a party hat on top of the globe, touching the globe along the chosen parallel, as Figure 2.1b shows.* The hat can, if we wish, intersect the globe by entering it along one circle and emerging from another circle below. No matter how we choose these circles, the resulting conic projection will resemble a cake—potentially one with a circle missing in the middle if the hat intersects the

*Usually, by the "top of the globe" we mean the North Pole, but nothing is preventing us from placing the tip of the hat above Hawaii or Zimbabwe.

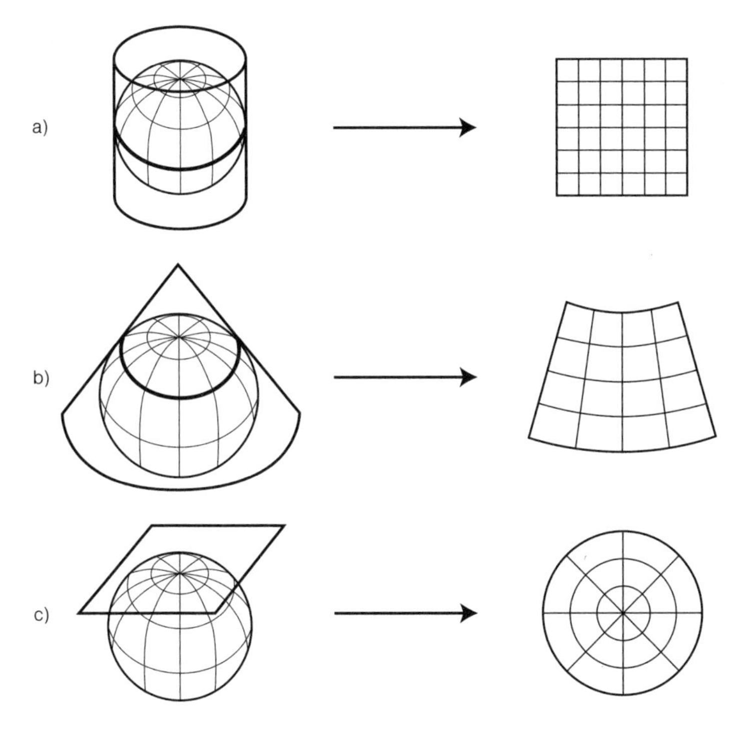

Figure 2.1 The three main types of projections: *a*) cylindrical, *b*) conic, and c) azimuthal.

globe—with a few slices missing. Assuming that the tip of the hat is above the North Pole, meridians will become straight lines radiating from the center of the cake, while the parallels will be marked as circular arcs centered in the middle of the cake. In this projection, the angles between meridians become smaller than the true angles.

Finally, we can simply keep the sheet of paper flat, so that it touches the globe at only one point—usually one of the poles (Figure 2.1c). The resulting azimuthal projection will be circular, with meridians radiating from the center, spaced at true angles. The parallels, on the other hand, will be marked as circles of different radii, centered at the pole. The name *azimuthal* stems from the Arabic word meaning "direction," and we'll shortly see why.

In high school, we usually stop our learning at these three types. But hundreds of different projections exist, and the bulb-in-the-middle-of-the-globe method, which gives us only three, with some slight variations, is just the starting point. After projecting the globe's surface onto a sheet of paper—or, rather, developing a system of equations describing such a projection—the cartographer adjusts the equations to get a map with desired properties. And that's where the fun begins.

Made for Navigation

Geert de Kremer was born in 1512 in Flanders, today's Belgium. Thanks to his uncle's connections, the poor shoemaker's son was able to attend a prestigious monastic school. He spent a lot of time studying Latin and Christian scriptures, but what changed the course of his career was the constant copying of sacred texts. He developed a particularly stylish italic script and became such an expert that, in his late twenties, he published a manual creatively titled *Literarum latinarum, quas Italicas cursoriasque vocant, scribendarum ratio* (*How to Write the Latin Letters Which They Call Italic or Cursive*). In the sixteenth century, the cool kids wanted a Latin name, so eighteen-year-old Geert enrolled at the University of Louvain as Gerardus Mercator,* where he attended lectures by Gemma Frisius,† the Dutch scientist we met when discussing triangulation. After graduating, Mercator didn't want to leave the charming city, so he persuaded Frisius to take him on as an apprentice. Together with the goldsmith Gaspar van der Heyden, they created objects ranging from globes to surgical equipment. Finally, Mercator could show off his excellent cursive by engraving labels on their collaborative creations.

Although he started as a calligrapher, Mercator quickly learned how to make globes himself, which was a task as hard as it was mundane. He had to meticulously craft twelve paper gores (triangular segments of the globe), carefully cut them out, and paste them onto a sphere. A single mistake—a slightly misplaced gore, for example—could destroy the whole project. Always eager to learn, Mercator soon started thinking about the opposite problem: instead of assembling a globe from flat pieces of paper, he wanted to make a flat map from the globe. This turned out to be harder to achieve than he originally thought.

* Mercator is the Latin for Flemish *kremer,* or English "merchant." It's even more complicated than that, as his family name has a few variations, such as Cremer or Kramer, but I'll spare you the linguistic details.

† Born Jemme Reinerszoon—I told you that Latin was all the rage!

Mercator lived in a time of constant "discoveries" of faraway lands yet to be found by Europeans. He was born only fifteen years after Christopher Columbus reached the Americas, starting the shameful era of rapid colonization. European maps updated with newly sited lands might be the only positive result of this tragic period in world history. While colonists were busy decimating indigenous peoples, cartographers back in their homelands faced a Sisyphean task. They would painstakingly work on a map for months, only to realize that the most recent geographical discovery made their shiny new creation out of date.

New information about previously unknown lands was streaming in. However, the data quality left a lot to be desired. Each traveler seemed to report a different size, shape, and location of the land that he (it was almost always a "he") visited. It's one thing to constantly update the maps; it's another thing entirely to decide which data to trust. Mercator hypothesized where all these contradictory reports were coming from. Sailors were navigating the sea by setting the compass toward a specific direction—*If I keep a constant direction on the compass,* they thought, *I'll follow the shortest route.* Mercator received geographical data based on this intuitive but incorrect assumption. Having followed the work of a Portuguese mathematician, Pedro Nunes, he realized the sailors' mistake.

Nunes had heard about the doubts of his compatriot Martim Afonso de Sousa, who had ventured to Brazil to take control of the area before anyone else (as in, any European) got in there first. To get back to Lisbon from Rio de la Plata, de Sousa needed to head toward the northeast, but he struggled to establish the exact direction. Inspired by this conundrum, Nunes understood that a sailor can follow one of two reasonable trajectories: he can either take the shortest route, which is an arc of a great circle, or a route that always follows the same compass direction. He proved that, despite sailors' common belief, these two options are not the same, unless they want to sail along the equator or any meridian. He called the route of constant compass direction the rhumb line, today known as the *loxodrome,* from the Ancient Greek *loxós* ("oblique") and *drómos* ("running"). On the globe, rhumb lines generally are not great circles but spirals. This is because a ship following a constant compass direction crosses each meridian at the same angle, as in Figure 2.2.

Mercator became exasperated by sailors making the same mistakes and bringing back unreliable data, so he decided to make a map on which rhumb

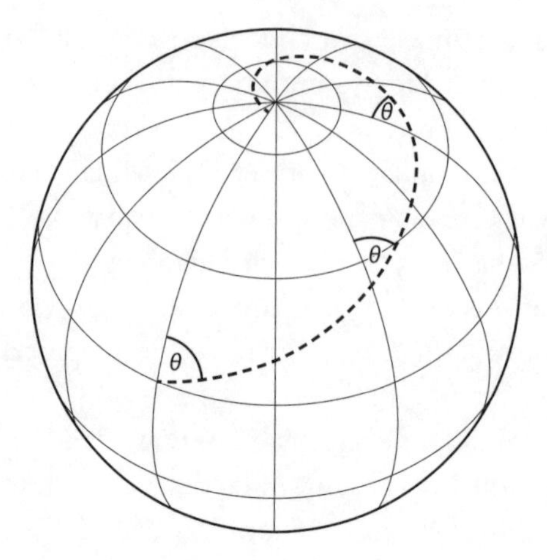

Figure 2.2 On the globe, a rhumb line crosses all meridians at the same angle, creating a spiral.

lines would correspond to straight lines instead of spirals. And that's how the famous Mercator projection was born.

He started from a cylindrical projection—placing an imaginary bulb in the center of an imaginary globe wrapped in an imaginary rectangular sheet of paper—and ended up with a rectangular map with parallels projected as horizontal straight lines and meridians projected as vertical straight lines, similar to the one in Figure 2.3. That's already a bit odd—on the globe, meridians meet at the poles, which is impossible for parallel lines. Indeed, on cylindrical projections, the poles stretch from single points on the globe to straight horizontal lines, like the parallels. This means that distances between meridians are constant on the map, while in reality, they get smaller and smaller as we travel from the equator toward the poles. On the other hand, the map depicts the distances between the parallels as if they are increasing for latitudes closer to the poles, while on the globe these distances remain constant. As we approach the poles, the distances on the map become so large—infinitely large, to be precise—that an Arctic snowflake should be depicted as bigger than Asia or Africa, so in practice we need to cut out the highest latitudes. Also, on the map, each parallel has the same length, which of course isn't true for circles of decreasing radii.

This simple cylindrical projection achieved one of Mercator's goals: parallels and meridians formed a grid of perpendicular straight lines. But he didn't stop there. He wanted the shapes of countries and continents to

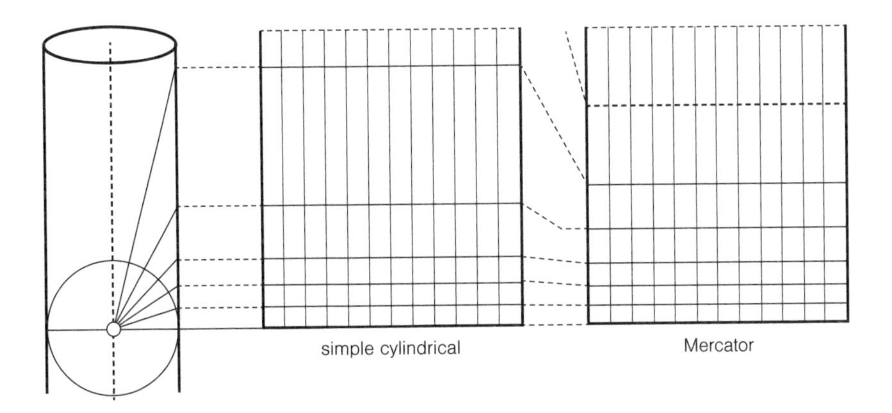

Figure 2.3 The Mercator projection is based on a simple cylindrical projection.

reflect reality—in other words, he wanted distortions in the north–south and east–west directions to be equal. To achieve this, he extended the distances between parallels on the map by as much as the distances between meridians were already stretched. Because the projection increased the gaps between meridians closer to the poles more than between meridians closer to the equator, the increase in the gaps between parallels also had to be bigger for latitudes closer to the poles than to the equator. By how much, Mercator couldn't know exactly, as the mathematical formula involved logarithms, derivatives, and integrals that hadn't been invented yet. Unfortunately, we don't know how he made this adjustment without the necessary tools; what we do know is that Mercator had great mathematical intuition and was fully aware of what he was doing. In the map's verbose legend, he wrote:

> In making this representation of the world we had . . . to spread on a plane the surface of the sphere in such a way that the positions of places shall correspond on all sides with each other both in so far as true direction and distance are concerned and as concerns correct longitudes and latitudes . . . With this intention we have had to employ a new proportion and a new arrangement of the meridians with reference to the parallels . . . It is for these reasons that we have progressively increased the degrees of latitude towards each pole in proportion to the lengthening of the parallels with reference to the equator.

Whatever he did, it must have worked, because in 1569, the first angle-preserving map was introduced to the world as *Nova et Aucta Orbis Terrae Descriptio ad Usum Navigantium Emendate Accommodata* (*A new and enlarged description of the Earth with corrections for use in navigation*). Not the catchiest title, but helpfully descriptive of exactly what he'd achieved. The map consisted of eighteen sheets, 33×40 cm (13×17 in.), that together with the border added up to a sizeable area of 202×124 cm (80×49 in.). These measurements made a beautiful wall decoration but not a practical navigation tool. Even smaller copies couldn't immediately make a useful addition to sailors' standard equipment, as some crucial technology was still missing. First, they didn't know how to precisely determine the longitude at sea. Second, compasses indicated magnetic directions, not geographical directions required by the map.* By the eighteenth century, these two problems were solved with the invention of the marine chronometer and a deeper understanding of the Earth's magnetic field. What helped Mercator's creation to fully take off, however, was math.

Math for Geography, Geography for Math

Having read about the principles behind the Mercator projection, you might think it's possible to go ahead and draw it. If only it was that simple. We know how to project the globe onto a cylinder, but the second step of the Mercator projection—the stretching of gaps between parallels—isn't clear. Mercator didn't say how exactly he did it, making it all but impossible to reproduce his result. Maybe that's a good thing, as the quest to mathematically describe the Mercator projection led to a surprising mathematical result.

In 1599, thirty years after the release of Mercator's map and five years after Mercator's death, an English mathematician, Edward Wright, published a detailed description of the mathematical principles of the Mercator projection. In *Certaine Errors in Navigation,* he tackled the mistakes made by sailors in navigating the sea. Since Mercator had designed his projection

*The geographic (true) north is the direction of the geographic North Pole, which is the point where all meridians meet in the north. Magnetic north is the direction indicated by a compass needle, which aligns with the Earth's magnetic field. As opposed to geographic north, magnetic north changes over time, following changes in the Earth's magnetic field. In September 2019, the two norths aligned for the first time in over 360 years.

to prevent these errors, it's only natural that Wright focused his research on understanding this map's mathematical principles. Recognizing the potential of the Mercator projection to aid sailors, Wright aimed to develop exact formulas for its vertical and horizontal stretch.

A good grasp of today's high school trigonometry should suffice to understand the horizontal stretch, as described by V. Frederick Rickey and Philip Tuchinsky in their 1980 article "An Application of Geography to Mathematics: History of the Integral of the Secant." Imagine the Earth as a peeled orange where one segment corresponds to a wedge of the Earth, as in Figure 2.4. If you slice this segment horizontally in half, you'll get a vaguely triangular ABC with two equal sides measuring the length of the Earth's radius (AC and BC) and one arc of the equator (AB). If you slice the segment a second time, but now cut above the first horizontal slice, you'll get the same shape, but smaller, corresponding to some parallel other than the equator (MNP). On Mercator's map, the projected arcs AB and MN have equal lengths, which means that to find the horizontal stretch, one needs to compute the proportion between the real-world AB and MN. By marking a few angles and applying simple trigonometry, we can find that this proportion is the secant of the latitude θ, or $\sec(\theta)$, which is just a fancy name for the length of NC divided by the length of NP in Figure 2.4.* So, a projection of an arc of latitude θ whose real-world length is L will have the length $L \times \sec(\theta)$.

Finding the length of a projected arc of a meridian turned out to be way more complicated and required the tools of calculus, not yet invented, but this unimportant detail didn't scare Wright—and it won't scare us. First, we need to understand how angles depend on lengths. For example, if we imagine a triangle, we notice that doubling one of the sides will change the angles; to keep the angles constant, we must double the other sides too. In general, if we stretch a figure by the same factor both horizontally and vertically, its angles won't change, which is the key property of the Mercator projection. We already know that the horizontal stretch depends only on the latitude, whose value will change as we move along the meridian. How can we ensure that the map is stretched horizontally and vertically by the same factor if this factor changes as we move vertically?

*By definition, $\sec(\theta) = \mathrm{NC}/\mathrm{NP}$. Because NC and BC both have lengths equal to the Earth's radius, this means that $\sec(\theta) = \mathrm{BC}/\mathrm{NP}$. Finally, since the "triangles" MNP and ABC are similar, the ratio BC / NP is equal to the ratio AB / MN. So, $\sec(\theta) = \mathrm{AB}/\mathrm{MN}$.

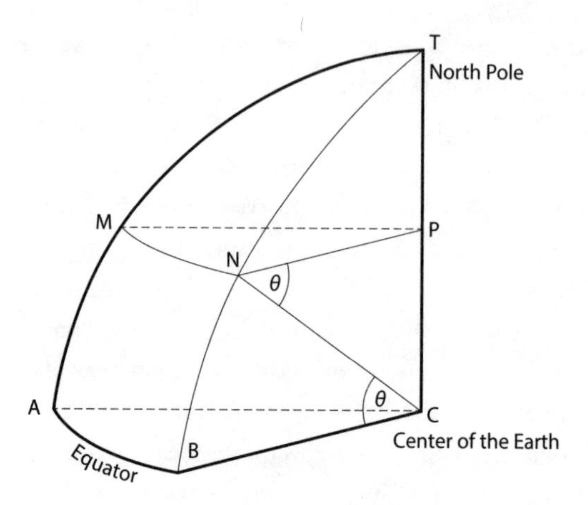

Figure 2.4 A halved wedge of the Earth as described by V. Frederick Rickey and Philip Tuchinsky.

For a small vertical arc, like l in Figure 2.5, the difference between the upper latitude θ_1 and the lower latitude θ_0 is small, so we can safely say that $\sec(\theta_0)$ and $\sec(\theta_1)$ have almost equal values and l's projection has length $l \times \sec(\theta_1)$ to match the horizontal stretch. To find the projected length of a longer vertical arc, we can divide this arc into many tiny segments, find the projected length of each, and then add them all up. As Figure 2.5 shows, for example, the projected length of the arc L will be $l \times \sec(\theta_1) + l \times \sec(\theta_2) + l \times \sec(\theta_3)$. And that's what Wright did. To compute the projected length of any vertical arc, he divided it into many tiny arcs. Then, he multiplied the length of each little arc by the secant of its uppermost point's latitude, and added them all up, calling this procedure a "perpetuall [sic] addition of the Secantes." To ease the navigator's task of calculating these lengths, he published a convenient table of already computed values for latitudes lower than 75 degrees.

In case this feels like too much math, Wright offered an alternative, physical interpretation of the Mercator projection. This time, it might be helpful to imagine the Earth as a spherical balloon wrapped in a cylinder as earlier, when the globe had a lightbulb inside. Now blow up the balloon so that it expands as much as possible,* and mark every point on the balloon that touches the cylinder. Unwrap the cylinder, and voilà, your Mercator projection is ready!

* Balloons, like calculus, were not invented yet, so Wright wanted the reader to imagine the globe "swal like a bladder." I thought you might prefer the balloon analogy.

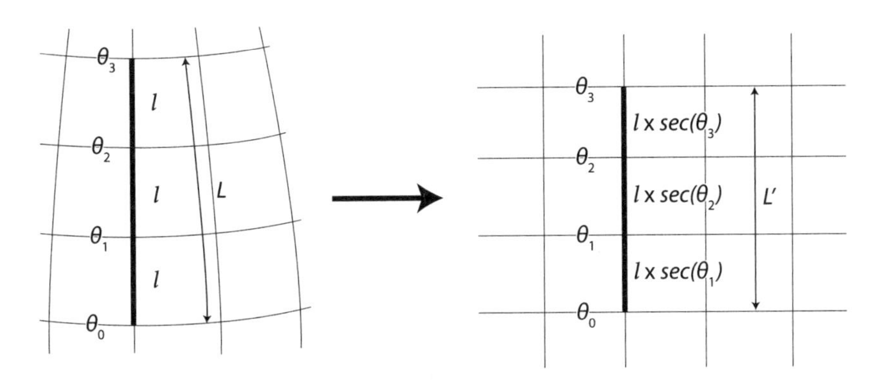

Figure 2.5 If l is very small, we can find the distance L' on the Mercator projection as a sum $l \times \sec(\theta_1) + l \times \sec(\theta_2) + l \times \sec(\theta_3)$.

Once again, mathematics came to the aid of cartography and navigation, and perhaps the story could end here. But geography was able to reciprocate mathematics with an ingenious insight.

In the early seventeenth century, a mathematician named John Napier published a table of logarithms,* including logarithms of sines† often used by astronomers. Yet another English mathematician and astronomer, Edmund Gunter, quickly followed with a table of logarithms of tangents.‡ When Henry Bond, a self-described "teacher of navigation, survey and other parts of the mathematics," compared Gunter's table with Wright's tabulated distortions on the Mercator projection, he noticed a stunning similarity. This inspired him to suggest the value of an integral commonly used to torture calculus students, including the author of these words:

$$\int_0^{\theta} \sec(x)\, dx = \ln\left|\tan\left(\frac{\theta}{2} + \frac{\pi}{4}\right)\right|.$$

*Don't worry if you've forgotten what logarithms are. Before the age of computers, logarithms were an astronomer's best friend, as they turned difficult multiplications into easy additions. Instead of painstakingly multiplying two large numbers, an astronomer would find the logarithms of these numbers in tables, add them up, and undo the logarithm of the result, again using the tables. It sounds complicated, but it's much easier and faster than multiplication.

†Like secant, the sine is a trigonometric function. In a right triangle, we find the sine of an acute angle by dividing the length of the side opposite to this angle by the length of the longest side of the triangle.

‡The tangent also is a trigonometric function. In a right triangle, we find the tangent of an acute angle by dividing the length of the side opposite to this angle by the length of the side adjacent to this angle that isn't the triangle's longest side.

The daunting expression on the left-hand side of the equality is a modern way of expressing Wright's sums of secants. He intuitively knew to calculate a larger distance by summing up many tiny distances, a method that later became the basis of calculus. Calculus is all about taking many tiny steps and thinking about what happens when "many" becomes "infinitely many" and "tiny" becomes "infinitely tiny." While Wright expressed the distance between the equator and the latitude θ on the Mercator projection as a sum of secants, today we'd say it's their integral. Integrals are sums of infinitely many infinitely small objects, which we denote with the elongated letter S (from *summa,* Latin for "sum"), like in the left-hand side of the equality.

This equality between the integral on the left and the logarithm on the right took over two decades to prove, despite attempts by the most skilled mathematicians, and was cracked only under the pen of a Scottish mathematician and astronomer named James Gregory. And that is how the quest to solve a problem in cartography using mathematics helped us to find the key to a tricky mathematical puzzle.

These mathematical developments popularized Mercator's creation. Not only did it change navigation forever, but it also became the standard world map projection for educational purposes. We use it even today, for better or for worse.

Size Matters

The Mercator projection preserves the angles, local directions, and shapes of lands. Thanks to Gauss, the most famous cartographical killjoy, we know that something must give: the sizes.

Around the equator, everything looks much as it does on the globe, but as we move toward the poles, weird things start to happen. The meridians normally meet at the poles, but cylindrical projections make them parallel, increasing the distances between them. This stretches the shapes horizontally, like the mirrors at carnivals that expand us. In addition, Mercator increased the gaps between parallels, which also stretched everything vertically, like those carnival mirrors that make us tall and thin. He ended up with shapes stretched in both directions, which at least made the deformation proportional. In other words, if we drew a small circle around any point on the globe, like in Figure 2.6, it would still be a circle, and not an oval, on the map. But the proportions are kept only between and not within directions,

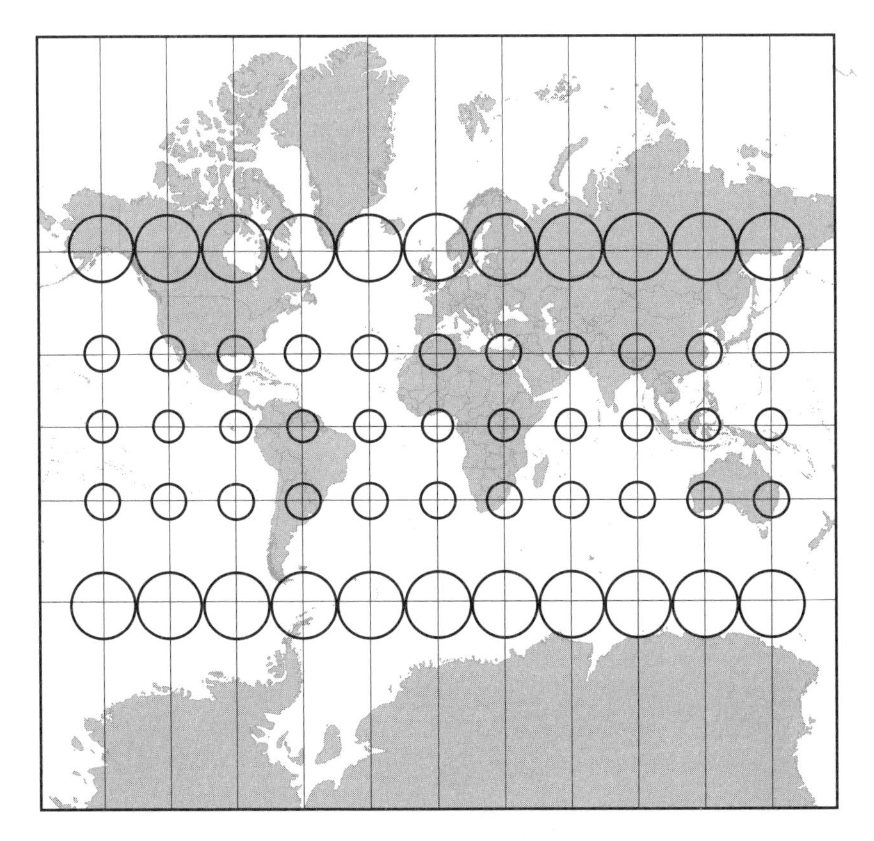

Figure 2.6 Each small area on the Mercator projection is distorted proportionally in all directions, which preserves the shapes of continents. The distortions are larger for points closer to the poles.

which means that two circles of the same radius on the globe might have different radii on the map. In particular, circles close to the poles would become much larger than circles close to the equator, despite them having the same size on the globe.

So, the distortion is more pronounced closer to the poles than to the equator. Way more. While the projected equator is the length it should be, the parallel of latitude sixty degrees (passing close to Helsinki, the Kamchatka Peninsula, and Shetland) is twice as long, seventy-five degrees (Greenland) fifteen times as long, and eighty degrees (the Norwegian archipelago called Svalbard) thirty-three times as long.* Taking into account

* I'm mentioning only the parallels in the northern hemisphere because in the southern hemisphere, they don't pass by anything but the ocean and Antarctica.

both the horizontal and the vertical stretch, the result is sizeable area distortions in regions that are far from the equator.

This distortion conveniently follows the common worldview of the powerful North and the insignificant South. According to the Mercator projection, the whole of South America has a size comparable to Greenland, when in reality it's over eight times bigger. And, in reality, the seemingly huge Alaska is smaller than Mercator's inconspicuous Libya, and Africa could fit the whole contiguous USA, India, Japan, China, and a large part of Europe within its boundaries. The world isn't as it seems on the Mercator projection.

Seeing these distortions, it's easy to call out Mercator as Eurocentric, if not racist, but before jumping to conclusions let's recall the mathematics behind his projection. By construction, the parallels of middle and polar latitudes in cylindrical projections stretch out horizontally. Mercator chose to also increase distances vertically for purely practical reasons: he created his map to solve the problem of sea navigation. For sailors, land areas are less important than knowing the angles between points. At sea, it doesn't matter how long it takes to get to a destination if we don't get there at all!

Having said that, I suspect that Mercator welcomed these distortions—though, again, more for practical than political reasons. When he was creating the projection, the age of land "discoveries" had just begun. To draw a map, the cartographer needs to know the shapes and sizes of lands, and Mercator had spent years collecting as many reports and sketches from expeditions as he could get his hands on. One of the sources to whom he regularly wrote was the English mathematician and astrologer John Dee, who had become a close friend of Mercator when they'd studied together in Louvain. Dee served as a scientific, political, and astrological advisor (a true Renaissance man) to both Queen Mary and Queen Elizabeth I.* He championed the colonization of the Americas and allegedly even came up with the term "British Empire." Dee happened to have one of the largest libraries in England and a great many connections and was able to provide Mercator with information invaluable for map creation. Mercator also

*At one point, Dee was also imprisoned for magical activity, such was the variety of his interests. In 1604, he wrote a letter to James I, asking the king to refrain from accusing him of being "Conjurer, or Caller, or Invocator of Divels, or damned Spirites." In the letter, he promised that he was innocent and ascertained that a person committing such a crime would need "to be stoned to death; or to be buried quicke: or to be burned unmercifully." It is often believed that Dee inspired Shakespeare's character Prospero in *The Tempest.*

received some help from his son Rumold, who was working for a London bookseller. Thanks to his networking skills and the forging of many connections of his own, Mercator managed to gather enough information to create his famous map.

Inevitably, then, Mercator knew much more about Europe than any other place in the world. It would have made sense to him that his home continent took up disproportionately more space on the map, allowing him to squeeze in all the little places he knew about. This way, the vast undiscovered areas in Africa didn't stand out so much in comparison to the well-described Europe of seemingly comparable size.

Mercator never intended his map for classroom walls or textbooks. He had one clear goal: to help sailors navigate the sea. But the map took on a life of its own. Publishers embraced the convenience and aesthetics of a rectangular world map with right angles that neatly fit into an atlas or onto the wall. Centuries after its creation, Mercator's map became *the* world map. It is the map that shaped my worldview during geography classes—and it likely shaped yours, too.

"Ragged, Long, Winter Underwear"

In May 1973, journalists attending a press conference in Bonn found themselves with copies of a new map in their hands, the same map that shocked Boston students over four decades later. The star of the conference was Arno Peters, the map's author—or, I should say, self-proclaimed author.

Peters's road to cartography was as atypical as it gets. Born in 1916 to a German family of social activists, from an early age he was exposed to ideas of equality and social justice. As a teenager, he spent most of his time doing sports, with some success in cycling and swimming. But it was cinematography that he decided to pursue as a career, so he headed to Hollywood to study American filmmaking. As a twenty-five-year-old, with his filmed production of a musical *Immer Nur Du,* he was one of the youngest filmmakers in Germany. In 1945, he received a PhD from Friedrich-Wilhelm University in Berlin, after defending a particularly timely dissertation, "Film as a Means of Public Leadership," but his interests were slowly shifting from film production to history and social studies.

Disappointed by the disproportionate representation of European civilizations in world history, he spent a large part of his academic career studying ways to represent all eras and all areas equally. He developed a "synchronoptic world history" timeline from 1000 BCE to 1952 CE, which gave equal weight to all the world's territories. This timeline was a visual representation ("optic") of events that happened at the same time in different places ("synchronously"). Peters believed that such representation would help us to better understand world history.

During his research, he consulted a large variety of maps, which were mostly biased toward the European point of view. None of them aligned with Peters's goal of equal representation, so he developed a new projection and started an aggressive campaign to replace maps based on the Mercator projection with his "Orthogonal Map of the World." Peters, a charismatic and skilled marketer, presented this map as the solution to the world's problems, including "the Europe-centered nature of our geographical picture of the world and its conquest." This lengthy and not particularly catchy phrase became the title of a brochure accompanying Peters's map.

Similarly to Mercator, Peters used a cylinder to project the globe. This time the cylinder didn't wrap the globe but cut through it at latitudes of forty-five degrees north and south. When the globe was projected, Peters doubled the length of the cylinder, thus stretching the map vertically by a factor independent of the latitude, unlike Mercator, whose stretches increased toward the poles. This resulted in a projection that preserves the relative sizes of landmasses. On Peters's map, South America is over eight times bigger than Greenland, Alaska is smaller than Libya, and Africa seems enormous, as it should. But this "new" map came with new problems—starting with its authorship.

Peters failed to mention that a strikingly similar projection had already been described a century earlier by Scottish clergyman James Gall. Ignoring Gall's creation, Peters repeatedly misled the public by claiming that he had created the first equal-area map. But Gall's projection hadn't been the first equal-area projection either. In 1772, a Swiss mathematician, Johann Heinrich Lambert, had published descriptions of a bunch of different equal-area projections. More recently, in 1910 a German geographer, Walter Behrmann, had described another equal-area projection (in my humble opinion, much more aesthetically pleasing). So, Peters's projection wasn't as revolutionary as he claimed, but it is still worth the hype.

While Mercator sacrificed areas for angles and shapes, Peters focused on presenting correct areas of countries and continents, in line with his quest to abandon the Eurocentric worldview. The rich North lost its dominance over African and South American nations—if only on the map. This, unsurprisingly, got the attention of various developmental organizations. In 1983, UNICEF adopted the Gall–Peters projection, and in the United Kingdom the map's popularity grew, mostly due to its use by the UK relief and development agency Christian Aid: it became a symbol of the fight for equality.

This approval by humanitarian organizations wasn't reflected in academic circles, however, with some cartographers criticizing its shape distortion. Just as images of yourself in two different carnival mirrors—one that makes you tall and skinny, the other one short and wide—might have the same area but different shapes, the areas in the Gall–Peters projection are in the same proportions as on the globe, but their shapes are skewed. The esteemed cartographer Arthur H. Robinson compared landmasses on this projection to "ragged, long, winter underwear hung out to dry on the Arctic circle," and it's hard not to agree. The Gall–Peters continents are long and drooping and look like deformed versions of the images we see in pictures of the Earth taken from space. The north–south proportion is accurate, at least in size, but the Gall–Peters map won't teach us much about the shapes of countries and continents.

Worldviews

So, why did the administrators of Boston Public Schools (BPS) decide to provide their social studies teachers with maps based on the Gall–Peters projection? This 125-school district is among the most diverse in the United States: 86 percent of the district's over 50,000 students are nonwhite, and English is the first language of about only one in two students. These statistics pushed the schools to think about and include nonwhite perspectives in the curriculum. Although the Mercator projection wasn't created to propagate racist views of the world, it has been misinterpreted as such, and comparing these two different projections can be a useful exercise for students and teachers alike. Soon after the introduction of the new maps, Natacha Scott, BPS's then-director of history and social studies, told *The Guardian* that "it was amazingly interesting to see [the students] questioning what they thought they knew." Every teacher should try to show their students different perspectives; as long as the Gall–Peters projection doesn't substitute but goes alongside the Mercator projection, it could indeed be a valuable addition to the curriculum.

The best solution, in my opinion, is to show students a variety of projections, so that they can see for themselves how different ways of representing the globe on paper create different distortions. Some activists even go as far as to advocate for ditching maps altogether and teaching geography using its most accurate representation: the globe. I strongly disagree, for two reasons. First, it's not practical; while maps fit snugly in students' backpacks and don't cost much to print, globes take up much more space and are less affordable. Second, children (and later adults) will be exposed to two-dimensional maps, whether we want it or not. They must be aware of the distortions, so they can then become critical map users. Globes should accompany—not replace—maps in education.

The question arises as to whether map projections impact our thinking in any significant way. Opinions are divided. In the nineties, researchers from the University of Arizona studied how this exposure impacted students' worldviews. The authors analyzed 438 maps drawn by first-year geography students from twenty-two cities. The participants had been asked to spend about half an hour sketching the world political map from memory. In fifteen participating cities, students exaggerated the size of their home continent, which doesn't surprise me—despite my awareness of map distortions, I still half-consciously think of Europe as larger than it really is. Strikingly, the seven exceptions came from students in all five African cities participating in the survey, as well as from Venezuela and Australia. It seems (although we'd need more studies to be sure) that the impact of the Mercator projection on students' worldviews trumps the tendency to exaggerate the size of our homelands. In particular, samples from all twenty-two cities contained maps depicting Europe as larger than it really is. Europe was the most consistently exaggerated continent in almost all countries apart from Kuwait, where Asia's size was overestimated to an even greater extent. On the other hand, students in most cities sketched a diminished Africa. Only Moroccans and, for some reason, Arizonans seemed to realize the true vastness of this continent. Also, based on the number of countries correctly marked on maps, students' knowledge of African countries was limited. All in all, the Eurocentric worldview prevailed.

However, a more recent study published in 2020 by Belgian researchers from Ghent University didn't find the "Mercator effect." The authors designed a simple online game and asked almost 100,000 users from all over the world to play it. Of course, I couldn't help but try it out, and I encourage you to do so too.* When I opened the page, I was shown a contour of South America above a contour of Europe. My task was to adjust their relative sizes so that they reflected the true relative sizes of these continents. Aware that Europe is smaller than I think, I kept shrinking it down—but by how much? I couldn't decide! At some point, I was satisfied with the result, so I clicked "OK" and moved on to the next pair: Turkey and the United States. Again, I was stumped—I knew that Turkey is smaller, but how much smaller? And so I struggled with all ten pairs of territories, embarrassed by my ignorance. At the end, I was shown four map projections (including the Mercator and the Gall–Peters) and asked to choose the one I was most familiar with. Finally, after answering a few demographic questions, the dreaded results appeared. Without revealing the number, I'll just say that if you think you did badly, I didn't do much better, despite being the author of a chapter on map projections. This just goes to show that familiarity with different projections and the awareness of their distortions don't guarantee that we can accurately estimate size proportions.

That's what the researchers concluded. The results indicated that how well we understand true size proportions has more to do with the actual ratio between these areas than with the map projection we're most familiar with. Small countries tended to be overestimated and large ones underestimated, or at least overestimated to a lesser degree. This didn't change even for elderly participants, who are more likely than anyone else to have been exposed to the Mercator projection as children, thus providing another piece of evidence against the "Mercator effect." If the Mercator projection impacted our ability to compare continents' areas, we'd expect users to overestimate Europe and North America, and underestimate Africa, but only the first of these expectations was confirmed. Researchers suggested that this could be related to the Eurocentric worldview, as depicted in the media and in school curriculums, rather than map projections. Did familiarity with any of the four projections indicate more correct estimations? In this respect, the study compared only nineteen- to twenty-five-year-old men with a high school diploma, so the results are rather limited. In this narrow group, the ability to compare relative sizes of territories didn't depend strongly on the projection the subject was most familiar with.

* You can access the game at http://www.maps.ugent.be/.

We'd need more research to confirm or dispute whether a particular map projection can impact the way we perceive the world. What we know for sure, however, is that no single map can teach us about all aspects of the world's geography: some will help us to understand the relative sizes, some will help us with shapes, and others with directions. That's why the walls in my apartment are decorated with a variety of maps, to keep reminding me that no single map represents the truth.

Centuries Later

This discussion about the benefits and flaws of the Mercator projection isn't merely an academic matter interesting only to geography nerds. The map you probably rely on in everyday life, be it Google Maps, Apple Maps, or any other major online map provider, is based on this sixteenth-century invention. Issues around the Mercator projection impact us in the present as much as they did in the past.

Launched in 2005, the first version of Google Maps used a projection that didn't preserve the angles on a map. While it didn't bother users close to the equator, it annoyed anyone trying to navigate at higher latitudes. Streets in Stockholm, for example, didn't meet at correct angles, which made the directions less precise. Because back in the day most of Google's income came from richer phone and computer owners in the geographic north, the technology giant quickly turned to an old and tested friend, the Mercator projection. For small-scale maps, Google engineers used the formulas that describe the Mercator projection developed centuries ago, while for larger scales they implemented a slight variation. One can notice the difference between the "true" Mercator and this new projection—appropriately named Web Mercator—only at local levels. We, as everyday casual users, don't need to worry about this discrepancy, but it can become a problem when data transferred from one projection to the other end up twenty miles from the intended point. For this reason, some governmental agencies discourage using Web Mercator. And rightly so, because it's intended to help navigate the streets, not plan a missile launch.

While it's not known who implemented the first version of Web Mercator, it was Google who popularized it. It might be the only company to have given a memorable identifier to a projection, albeit unofficially. All projections have dull, number-based codes in the EPSG Geodetic Parameter

Dataset,* a public registry of all things geographical. Initially, the committee responsible for the registry refused to include Web Mercator in their dataset, citing its technical flaws. Then a blogger and maps fan, Christopher Schmidt, named it 900913, which is "Google" transliterated to numbers. It stuck, despite the committee finally assigning the official identifier EPSG:3785, which was later changed to the confusingly similar EPSG:3857.

Projections similar to Web Mercator could have been introduced as early as the 1980s when the first CD-ROM zoomable maps appeared, although probably the most famous of them—Microsoft Streets 98—went for a completely different projection. Also, the first online maps, such as MapQuest, introduced in 1996, seemed to choose projections rather haphazardly. It all changed with the success of Google Maps, dreamed up by Danish brothers Lars and Jens Eilstrup Rasmussen, a computer scientist and a software engineer, respectively, together with Sydney-based software engineers Noel Gordon and Stephen Ma. Following their success, other online map providers quickly adopted versions of Web Mercator. MapQuest, for example, admitted that Web Mercator is "more user-friendly, especially for those dealing with multiple data sources."

Web Mercator has the same problems as the original sixteenth-century projection. It distorts sizes in high latitudes, which is why maps need to be cropped at 85 degrees north and south and, as on paper maps, Greenland grows to the size of Africa. It took Google a decade to fix these problems and, in August 2018, they announced a big change to their maps. Now, when you zoom all the way out, you don't see a flat Mercator-based map with its skewed sizes, but a three-dimensional globe. We now have the best of both worlds—precise navigation at a local level and a great educational tool on a global scale.

Make It Pretty

In the 1960s, the American publishing and tech company Rand McNally asked Arthur H. Robinson, the prominent cartographer from the University of Wisconsin mentioned earlier in this chapter, to construct a new world

*The EPSG Geodetic Parameter Dataset was maintained by an informal scientific organization, the European Petroleum Survey Group, between 1986 and 2005. In 2005, the registry was taken over by the newly founded IOGP (International Association of Oil and Gas Producers) Geomatics Committee.

map projection useful both for wall maps and smaller-scale maps in atlases. Among other requirements, the map was requested to be aesthetically pleasing, the latitude–longitude grid "simple and straightforward," and the oceans uninterrupted. The latter condition refers to the fact that while we can turn the globe continuously, a flat map must be interrupted at some point. For example, on the Mercator projection, the Pacific Ocean disappears on the right and reappears on the left. Some projections (by the way, there are hundreds of named projections), such as Mollweide or Goode's Homolosine, tear the oceans apart in more than one place, which is what Robinson was asked to avoid.

So far, we've looked at lightbulb projections, which we can describe with mathematical formulas that take the longitude and latitude of a point and return two-dimensional coordinates of this point's projection on a plane. While these formulas look complicated and unintuitive, the idea behind them is no different than other math functions.

The first functions we learn about at school are linear, such as $y=x+2$. This formula tells us what to do to any number x: by adding two, we transform it into a bigger number, y. So, 2 becomes 4, –1 becomes 1, and 0.3 becomes 2.3. This simple formula changes one number into another, without changing the dimension. We could create a similar formula to replace a two-dimensional point (a pair of numbers) with a one-dimensional number. For example, we can say that each point (x, y) becomes $x+y$, and this way, we drop a dimension. Similarly, we can use the language of math to describe how to take any three-dimensional point* from the globe and transform it into a two-dimensional point on a flat map. In practice, instead of wrapping illuminated globes in sheets of paper, cartographers develop such mathematical formulas. A formula tells them (or the computer program) exactly where each point from the globe should land on a map. A bunch of equations that look abstract and confusing turn into a clear picture when plotted—the final look of the map is a mystery until this last step.

But there are exceptions.

*Because we can identify any point on the globe just by its longitude and latitude, it might seem that the globe is a two-dimensional space. There is, however, one extra number necessary to identify this point: the Earth's radius. Because we assume that the radius is identical for all points, we usually don't mention it but, strictly speaking, to identify each point we need exactly three numbers: the latitude, the longitude, and the radius.

Many items on Robinson's list of specifications were subjective, as they referred more to the map's appearance than its mathematical properties. How do you capture "simple and straightforward" in a mathematical formula, for example? Robinson quickly realized that he needed to take a nontraditional approach. Instead of writing equations, he drew a map (or rather made the computer draw a map for him), looked at it, frowned and adjusted the program to improve the picture. The new map didn't satisfy him, so he changed a few more parameters and produced an improved map. He continued with this trial-and-error process until he was pleased with the result, and only then did he describe his projection mathematically.

Robinson turned the method of projecting the globe onto a map on its head. Traditionally, the cartographer first decides which equations to use—or, equivalently, how to wrap a sheet of paper around the globe and where to put the light source—and only then does she or he learn what the map looks like. Robinson, on the other hand, first looked for a satisfactory appearance, and only later did he worry about the mathematical description.

Rand McNally seemed happy with the result, and in 1965 they printed a wall map based on this new projection. According to an anecdote spread in cartographic circles, however, the public didn't appreciate this change of worldview (which would partially explain the continued success of Mercator), and Rand McNally downgraded Robinson's creation to school atlases,* where it stayed for over a decade. Luckily for Robinson, in 1988 his projection became the map of choice of the National Geographic Society for a whole decade, fulfilling every cartographer's dream. It was indeed a big deal, as *National Geographic* doesn't take its map selections lightly. A year earlier, in December 1987, their chief cartographer, John Garver, had invited the most prominent cartographers to the society's headquarters in Washington, DC. Having discussed several whole-world map projections, they all agreed that Robinson's creation was second to none. It took a year of preparations for his map's debut in *National Geographic,* the magazine known for its yellow spine and outstanding photography. Soon, over five hundred other magazines and newspapers followed the magazine's lead.

*I would argue that no cartographic publication is more important than school atlases. For many of us, this is our first encounter with maps, and it shapes our worldview forever. Unfortunately, it seems that few cartographers share my opinion, so this new use for Robinson's map was considered a downgrade.

What's so special about the Robinson projection? It preserves neither the area nor the shape of the continents. This map is all about compromises—it distorts everything, but only a little bit. The balance between reflecting true shapes of continents (as in the Mercator projection) and true relative sizes (as in the Gall–Peters projection) makes the Robinson map suitable for educational purposes. After all, Robinson had created it with exactly this goal in mind—he was more interested in appearance than in particular mathematical features.

While Arthur H. Robinson is the only person credited for this projection, some suspect that he worked in tandem with another cartographer: Barbara Bartz Petchenik. There are multiple reasons to think that this is the case. Robinson and Petchenik worked together on so many projects that it's hard to believe that she didn't participate in creating the projection that became Robinson's legacy. For example, they co-authored one of the classic cartography books, *The Nature of Maps,* about the philosophy of map understanding. Petchenik's research interests included map design, cognitive psychology, and education, so she could have provided invaluable insights for the creation of this unusual projection. Although there's no evidence she participated in the design of the Robinson projection, given that female cartographers (and scientists in general) tend to be forgotten, I'm tempted to believe that her male colleague at least consulted with her during the map design. Decades after Petchenik's death, her name is known to young map enthusiasts all over the world, as hundreds of children send in their drawings for the biannual Barbara Petchenik Children's World Map Drawing Competition. Given her passion for cartography education, I'm sure this would have made her proud.

Which Way to . . . ?

The Islamic Center of Washington took over a decade to create, from the initial idea to its opening in 1957. Not long before completion, Mohammed Kamil Abdul Rahim, the Egyptian ambassador in charge of the project, visited the building site. Responsible for the prayers of thousands of Muslims about to attend the new mosque, the first thing he did was consult his compass. He wasn't happy with the result. "It points so far north," he commented in disbelief.

To fulfill one of the five pillars of Islam, Muslims pray five times a day, facing the holy city of Mecca in Saudi Arabia. More precisely, they orient themselves toward the Kaaba, a small shrine that contains the sacred Black Stone of Mecca. Determining the correct direction to Mecca—called *qibla*—is the most important step in building any mosque. Washington, DC, located at the latitude 39 degrees north, is, without doubt, north of Mecca at 21 degrees north, so it's easy to understand the ambassador's worries. If you look at a world map, the most popular Mercator projection, let's say, you'll see that the mosque should face toward the southeast, not northeast.

Irwin S. Porter, one of the mosque's architects, had some explaining to do.

He noted that the compass was pointing toward the magnetic north, which isn't the same as the "true," geographic north.* This didn't seem to convince the visitor, who went on to describe the rather ruthless way they built mosques in his homeland. Apparently, builders were given a point in sight to orient the mosque toward. After completion, the royal engineers would survey the building and, if the orientation was wrong, the mosque would be destroyed and built again from scratch. No pressure.

After this visit, Porter, understandably worried, asked a National Geographic Society cartographer to double-check the orientation, just in case. Once he received an all-clear to proceed with the construction, Porter asked the cartographer how he would be able to convince the ambassador that his orientation of the mosque was correct. "Tell him to take a globe," the cartographer advised, "and two thumb tacks, one at Washington and one at Mecca, and to run a string between and he will see that that is the direction of Mecca." So, Porter packed up a globe, some tacks, and a string, and sent it to the ambassador. This simple prop must have convinced the Egyptian official because when journalists from *The Washington Daily News* called the embassy to question the mosque's orientation, the secretary reportedly responded: "It's very simple, anyone can demonstrate it. Just take a globe and two thumb tacks and . . ."

Establishing qibla isn't easy. First, what does "direction toward Mecca" even mean? On a flat surface, to find a direction from point A to point B, we draw a straight line. But we've already seen that geometry on a sphere is rather more complicated. How do we define a straight-line segment between two points on a flat sheet of paper? We can do it in two intuitive

* For a reminder of the difference between the two, check the footnote on p. 40.

ways. On the one hand, it's the shortest distance between these two points. On the other, it's a path along which we keep constant direction, or angle. Unfortunately, on our curved Earth, these two definitions aren't equivalent, which means that we can have two different definitions of a "straight" path from a location to Mecca.

We could interpret qibla as a rhumb line, so a path toward Mecca that crosses all meridians at the same angle. As we saw, on the Mercator projection this definition does indeed correspond to a straight line. We could also think about qibla as the shortest path toward Mecca, which—as we discussed—follows the great circle. Along this path, the compass will constantly show us a different direction. In this case, we interpret the direction as the angle at the start of our journey and call it the azimuth. Although this discrepancy has led to some heated discussions among Muslim scholars, the great circle method is generally considered the accepted qibla.

Map projections that we've seen so far represent qibla in a way that perplexed the ambassador of Egypt, and rightly so. The Mercator projection shows qibla as a curved line heading toward the northeast, rather than the intuitive southeast. James Ireland Craig, a Scottish mathematician working as a cartographer in Egypt at the turn of the twentieth century, observed the struggles of his Muslim colleagues attempting to determine where they should direct their prayers. Finding all existing projections unsuitable for locating qibla, he designed a new one, today known as the Craig retroazimuthal or Mecca projection.

On retroazimuthal projections, the directions (known as azimuths) from every place on a map to a chosen location are correct. To solve the problem of locating qibla, Craig picked Mecca as the center of his projection. In Figure 2.7 you can see that the resulting map looks, well, weird, to say the least—a bit like a long, droopy mustache—which renders it completely useless as a wall map. It was, however, created with a concrete goal in mind, and it serves that purpose well. In Craig's map, the azimuth from any point to Mecca is a straight line, as expected.

Craig's map isn't the only retroazimuthal projection. In 1929, John Hinks designed a map centered at Rugby in the county of Warwickshire, UK. Thanks to his projection, colonials all across the British Empire could communicate by directing their antennas at the radio station in Rugby. And Edgar N. Gilbert, who worked for Bell Telephone Laboratories, jokingly replaced Mecca with Wall Street, the mecca of the financial world. Ret-

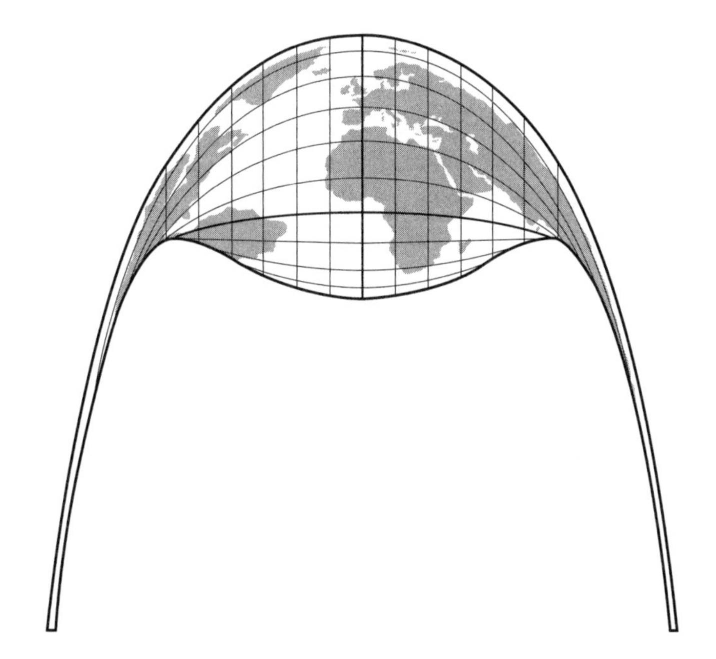

Figure 2.7 Craig retroazimuthal projection centered at (0° N, 0° E).

roazimuthal projections are a special type of azimuthal projection, on which directions *from,* not *to* a chosen point are correct. To understand the difference, take a globe and connect two points with the shortest path. The azimuth from A to B is the angle between the meridian of A and this path, while the azimuth from B to A is the angle between the meridian of B and this path. Figure 2.8 shows that these angles are not the same, which can be explained by the properties of the sphere. Points A and B together with the North Pole form a triangle. As we saw in the previous chapter, angles in spherical triangles—as opposed to flat triangles—don't add up to 180 degrees. Even worse, their sum isn't constant but depends on the latitudes of the two points. This means that there's no simple relationship between the angles in this triangle, one of which is the azimuth from A to B and the other one the azimuth from B to A, unless both points happen to have the same latitude.

While retroazimuthal projections help us find a direction *to* a specific location, such as Mecca, we use azimuthal projections when we care about directions *from* one chosen point. Sometimes, the appropriate use of an azimuthal projection might become a matter of life and death.

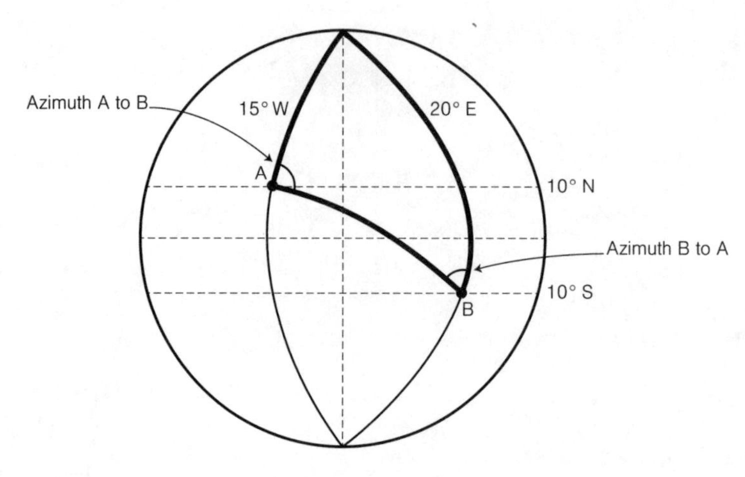

Figure 2.8 The azimuth from A to B isn't the same as the azimuth from B to A.

You Will See What I Want You to See

In 2003, *The Economist* published an article on the North Korean threat. The text was accompanied by a map of the estimated missile range, represented by concentric circles, which reassured European, American, and African readers—the majority of its subscribers—that they didn't need to worry about a potential attack from North Korea. According to that map, while a North Korean missile would endanger Asians and Australians, it wouldn't reach other parts of the world. Except that's not what one could conclude from the text of *The Economist*'s article, which claimed that the missiles could travel as far as 15,000 km from North Korea. Given that a missile could be launched in any direction, this would mean that nobody within a circle with a 30,000 km diameter could relax, which didn't match the story told by the accompanying map. Taking into account the Earth's circumference of about 40,000 km, some relatively simple math* shows us that a North Korean missile would put most of the world in danger. Something didn't add up.

The problem was, once again, an unfortunate choice of projection. The journalist picked the Mercator projection, which not only distorts the dis-

*We need to divide the area of a circle of radius $r = 15{,}000$ km on a sphere of radius $R = 6{,}400$ km by the sphere's surface area. In other words, we need to divide $2\pi R^2(1 - \cos(r/R))$ by $4\pi R^2$, which gives a proportion of the Earth's area endangered by North Korean missiles of about 85 percent.

tances close to the poles, it also makes us forget that the easiest way to reach North America from North Korea isn't to launch a missile to the east, through the Pacific, but to the north, above the Arctic. Readers quickly pointed out this geographical mistake, and *The Economist* corrected their blunder, thus destroying our peace of mind.

The North Korean threat became a hot topic again after UN sanctions were put in place in 2013 following nuclear testing. In 2015, the BBC published an article about the danger of a missile attack on the United States. While the accompanying map wasn't strictly wrong, it was certainly misleading. The BBC chose a type of azimuthal projection called orthographic, centered on North Korea, which distorted distances from the potential missile launch site to the target, especially near the map's edges. Able to show only one hemisphere, it also hid a sizeable portion of the United States—a rather peculiar choice, given the main topic of the article was a potential strike on that nation. The message could have been improved by using an azimuthal equidistant projection, like the one in the UN flag, which presents the entire United States and preserves true distances.

Sometimes, the choice of projection is misleading because of the author's genuine mistake. At other times, misleading projections are chosen deliberately, usually for political reasons.

Never was the power of projections more used and abused than during the Cold War. On the Mercator projection, the Soviet Union looks like a huge monster dominating the world—how convenient if you want to communicate the extent of the threat posed by the communist land. Add some red color to mark their territory and you could convince the biggest skeptic that the USSR was a nation to be afraid of. Western Europe seemed more endangered by the USSR than the United States, as the Mercator projection presents America and Asia on opposite sides of the map. This apparent geographical separation of the Americans from the Soviets could suggest to Europeans that their American friends might not be too keen to engage in a potential conflict. So many messages in one simple map!

On the other hand, if you wanted to stress the importance of US participation in European wars, you'd use an azimuthal projection centered at the North Pole. This picture brings the United States closer to the Soviet Union as well as to Western Europe. Suddenly, the land that appears so far away on the Mercator projection *could* potentially threaten Americans with a nuclear attack. Similar maps were shown to Americans during the Second

World War to encourage support for US participation in the conflict. This wasn't a far-flung battle, the maps indicated, but something that was on our doorstep.

The Soviet threat wasn't only theoretical, of course. In August 1983 at JFK airport in New York, passengers boarded Korean Air Lines Flight 007 heading toward Seoul in South Korea. After takeoff, it made its way toward Anchorage, a common refueling stop, as we've seen. Following a brief break, the plane took off again, and the crew set the autopilot for the South Korean capital—or at least that's what they thought. Unbeknownst to the crew, the autopilot started to shift the planned course and entered Soviet territory. The Soviets didn't like what they were seeing. They sent a signal to the Korean crew, informing them about the intrusion. Then they fired warning shots, but the plane continued on the erroneous route. Soviet pilot Colonel Gennadi Osipovitch later reported: "My orders were to destroy the intruder. I fulfilled my mission." All 269 people on board died when the plane was shot down just off Sakhalin Island in today's Russia.

The details of the tragedy were revealed only in 1992, after the Cold War had ended. This doesn't mean, however, that the public didn't learn about the incident when it happened in 1983. But, predictably, the two sides of the conflict developed two separate narratives. On one side, the Soviets defended their right to protect their territory. On the other, the Americans presented the change of flight path as an innocent mistake. Both sides illustrated their version of the story with different projections.

American news media preferred maps that magnified Soviet territory, showing the United States as a victim of the enemy nation. They chose projections that distorted the angles, so one couldn't infer much about the discrepancy between the plane's planned and actual route. Was it a slight mistake, or did the Soviets have a reason for concern? On the American choice of map, it seems that the flight path didn't change much, so the Soviets' reaction was considered unreasonable. Other maps presented a completely different story. On the Mercator projection, the change of trajectory seems significant and deliberate, exactly what we'd expect to see if the plane was spying above Soviet territory. With this picture in mind, it's easier to justify the Soviets' decision to shoot, despite the tragedy of lives lost. One event, two maps, and two interpretations. That's the power of maps.

Stay Alert

The impossibility of creating a flat map without any distortions gives cartographers the power to impact our worldview by choosing an appropriate projection. Does the "rich" North dominate in size and power the "poor" South? Should we worry about North Korean missiles being able to reach us? Are the United States and Russia worlds apart, or are they next-door neighbors? Different maps will prompt different answers to these questions.

Now and then, someone announces which projection is best, trying to replace all maps with a new, or indeed old, invention. But no projection is right or wrong; rather, it might be suitable or unsuitable in a particular context. I hope that schools all over the world will expose students to a variety of projections, so that we bring up a generation of conscious map consumers. We would all benefit from spending some time thinking about what the map's author wants us to see and what she prefers to hide. This way, it's possible to form our own opinion on the topic the map is aiming to represent, and a little power is taken away from the cartographer and put back into our hands.

3

SCALED

How to Measure a Line

As a child, I could spend hours looking at doll's houses displayed in toy shops. Everything seemed so real to me—the tiny kitchen sink, the miniature sofa, the minuscule lamps—just on a much smaller scale. I was amazed by how many details the artist could include despite everything being so small. I especially liked to compare a small and a large copy of the same doll's house, placed side by side. From afar, they looked identical, differing only in size (and price, as my parents noticed). After closer inspection, however, I could find differences. While the tiny plates in the bigger house had flowery patterns, the even tinier plates in the smaller house depicted only vague dots painted in similar colors. The family portrait displaying the features of the larger house's owners became a rough sketch in the smaller house. And, most disappointingly, while the dolls living in the larger house used the same brand of toothpaste as I did, the smaller house didn't even include toothpaste.

Doll's houses are models of real houses that belong to human beings. They have the same purpose: to provide a roof overhead and a pleasant, safe environment in which to live. But the ornaments and knickknacks that easily fit in a large house would make a smaller house too cluttered. When we reduce the scale, we must redesign some details, and maps are no different.

As with a small doll's house, a world map in a school atlas doesn't offer much space for details. The map's authors will have attempted to show as much of our vast planet as possible, without making it too cluttered to be useful. You won't be able to find your hometown, then, or maybe even your country. All but the largest cities, rivers, and mountain ranges disappear, and we are left with an overview of the world.

A large wall map of the world is a whole different story. Its authors will have had the freedom to fill up the space with many more details than was possible in the small school atlas. It fits all the world's countries, its large and medium cities, and many lakes and rivers. To discover every detail, you'd have to stare at this map for hours (which doesn't seem like such a bad way to spend a free afternoon). Cartography is the science—or art—of finding the balance between including as many details as possible and making the map readable.

Of course, the most accurate map would have the scale of a mile to a mile, as suggested by Mein Herr, the creation of the author of *Alice's Adventures in Wonderland,* the mathematician and photographer Lewis Carroll.* But such a map wouldn't be too useful, as "it would cover the whole country, and shut out the sunlight!," which, according to Mein Herr, worried the farmers. To make a map practical, we need to use a smaller scale.

At school, we learn a simple but not entirely correct definition of a scale: it's the ratio between the distance on a map and the real-life distance. Then, typically, we proceed to solve countless exercises asking us to convert the true distance to the map distance, and vice versa. The problem is that maps are two-dimensional representations of a three-dimensional globe, which means that to create a map we must distort some, if not all, distances. The curvature of the Earth makes the concept of a scale as we know it not entirely correct.

Still, all maps come with a numerical or graphical representation of a scale in the bottom corner—but how should we interpret it? Usually, it refers to the ratio between the size of the globe that was projected on the map and the size of the Earth. In practice, of course, what's projected onto a sheet of paper is not the Earth but its smaller representation. Returning to the analogy used in the previous chapter, we can think of the Earth as a massive (almost) spherical balloon with oceans and landmasses sketched on the surface. Before projecting these shapes onto a sheet of paper, we need to deflate the balloon until the size is satisfactory. Usually, the larger the area we want to represent, the smaller the balloon must become so that

* His real name was Charles Lutwidge Dodgson.

the whole area of interest fits onto the page after the projection. The ratio between the radius of the shrunken balloon and the initial balloon is called the *nominal scale,* which is what we most often see in the map legend—the little box displayed on every map that explains its features. When the map covers a relatively small area, such as a city, the impact of the Earth's curvature is negligible, and we can safely assume that the area is flat, at least in terms of the Earth's curvature. Then, the scale is almost constant for all distances on a map and, for all practical purposes, the school interpretation of a scale as the ratio of a distance on a map to the same distance in the real world is equivalent to the nominal scale.

The scale is usually presented in the form of "1: some number," where the larger the number is, the smaller the scale of the map. For example, let's compare a map on a scale 1:5,000 and 1:50,000,000. In the first one, one centimeter on a map will correspond to 5,000 centimeters, or 50 meters, in terrain. This means that everything you see around you will still be relatively large on a map. Such a scale is useful to depict walking distances in a city, for example, where the map user needs to know the names of roads and streets to navigate. One centimeter on the latter map, on the other hand, corresponds to 50,000,000 centimeters, or 500 kilometers, in the real world. This means that if you took a train from London to Edinburgh, during five hours of changing landscapes you'd cover only about one centimeter on this small-scale map.* The larger the map's scale, the more details it includes.

This is where the cartographer's skill comes in. It's up to the mapmaker which details to include and which to remove, so two maps of the same area on the same scale might differ a lot. As the scale of the map decreases, everything becomes more cluttered. To keep the map useful, the cartographer will need to reduce the amount of detail included in the process of cartographic generalization.

Some features might be omitted completely. Intuitively, we'd expect the smaller rivers, towns, or lakes to be removed first, and the larger ones to stay. This is, however, only partly the case, as the city of Baltimore, Maryland, has painfully experienced. Baltimore, with over 600,000 citizens, is by no means a small town, but it still disappears from many small-

*It might seem confusing that 1:50,000,000 is a smaller scale than 1:5,000. This is because the same object will look much smaller on a map with the former than with the latter scale.

scale maps, despite the proud presence of its smaller neighbors. It's not because cartographers particularly dislike this seaport city, known for the excellent Johns Hopkins Hospital and for legendary high crime rates. It's simply because cartographers don't like empty spaces, so they fill the surrounding available area with smaller towns but fail to squeeze in the nine letters of Baltimore. This tendency to optimize the use of space on maps, leaving the logic behind, is so common that it even has a name: the Baltimore phenomenon.

Other features are aggregated or merged. For example, individual buildings depicted on a large-scale map might get merged into an "urban area" when the scale gets smaller. Some can even be displaced. Brazzaville and Kinshasa, two African capital cities with long names on opposite banks of the Congo River, would overlap on small-scale maps, so cartographers sometimes move them slightly away from the river to fit in their names.

The most interesting distortion, however, happens to the lines on a map, such as borders and rivers.

Getting It Straight

Think of the map of the western United States. What do you see? Plenty of straight lines and rectangles. To understand why, we need to go back in time to 1607. Virginia was the first British colony in America, followed by the rest of the East Coast, from Maine to Georgia, which became the first thirteen English colonies. With a few exceptions, their borders didn't look particularly regular, as they mostly followed natural boundaries, such as rivers and mountain ranges.* This isn't unlike the rest of the world, where almost a quarter of all international borders follow rivers. But the western states that joined the Union later are a different story. Thomas Jefferson, the third president of the United States, is known for advocating for social equality. His most famous phrase, "all men are created equal," extended beyond condemning slavery (which, of course, didn't stop him from enslaving over six hundred people); he also suggested to Congress that all states should be of the same size and recommended dividing them according to the latitude–longitude grid. For example, the northwestern states were each

*Today's borders are modified, but they still mostly follow natural boundaries.

supposed to cover two degrees of latitude and four degrees of longitude. Congress wasn't too keen on Jefferson's proposal but ended up applying a similar method to draw quite a few borders, including the prairie states (the Dakotas, Kansas, and Nebraska) and the Rocky Mountain states (Colorado, Montana, and Wyoming).

Today, the borders of neighboring Wyoming, Utah, and Colorado seem to consist entirely of straight lines. Indeed, Congress defined Colorado's borders exactly like that: as a rectangle between the latitudes 37 and 41 degrees north, and between the longitudes 25 and 32 degrees west. But how rectangular is Colorado today? As we already know, the Earth is spherical, so Colorado's northern border must be shorter than the southern one, which means that it deviates slightly from a rectangle—and its borders aren't straight. Although the borders were defined as straight lines, it doesn't mean they became straight in reality.

In the nineteenth century, surveyors couldn't use satellites and precise measurement tools, so they had to do their best with magnetic compasses. They placed hundreds of boundary markers in lines as straight as possible so that they could establish the borders in a connect-the-dots fashion. While they did an amazing job with these limited tools, the lines aren't perfectly straight. Even the most precise nineteenth-century surveyor attempting to draw a straight line in the field would have made some tiny mistakes. In the case of Colorado, these tiny mistakes added up to hundreds of feet each, but these can be seen only on a large-scale map. Zoom in on your favorite online map and see, for example, how the southern border of Colorado turns south when it reaches Oklahoma, and then south again. Mistakes in this region are quite surprising, as this area is one of the flattest in the United States—one can't blame hills or valleys for disturbing the surveyors during their task. When we count all these deviations from the initially planned quadrangle (a polygon with four sides), it turns out that Colorado has 697 sides.* This is the power of the scale—the same objects on maps look one way when we zoom in and another way when we zoom out. In the case of Colorado and other US states, as we look closer, the seemingly straight lines turn into zigzags. And this makes measuring lengths on maps quite tricky.

*A polygon with 697 sides also has a (not very useful) name: hexahectaenneacontakaiheptagon.

An Accidental Discovery

Lewis Fry Richardson, a Quaker and pacifist, was a conscientious objector during the First World War, which didn't stop him from studying the causes of wars. Unlike most war researchers, he applied mathematics to answer his questions. He used, for example, systems of differential equations to describe interactions between countries.

In his book *Statistics of Deadly Quarrels,* Richardson gathered a great deal of data on conflicts resulting in human deaths, from murders to world wars—in other words, on deadly quarrels. He was looking for patterns in the occurrence of violent events and the number of victims, either temporal or spatial. While he didn't find any evidence for wars' dependence on time, he explored two contradictory theories on how conflicts depend on geography. Both theories assume that the length of the shared border between two countries impacts the likelihood of a war between them: neighbors either fight for resources or ally against other countries. Although the results of this investigation didn't make history, Richardson discovered something in the process—something possibly more important than the question he was trying to answer.

While analyzing the country border data, Richardson noticed many inconsistencies. Not the slight measurement discrepancies natural for any data-gathering process—no, these apparent mistakes were huge. For example, Portugal reported the length of their shared border with Spain to be 1,214 kilometers, while Spain estimated the same border as 987 kilometers. The difference is about the distance between London in the south and Sheffield in the north of England! Richardson found similar inconsistencies between reports on multiple borders in his dataset. So which report was right? Richardson realized that the answer was both. And neither.

To understand how it's possible, let's think about how measurements *should* behave. Imagine you need to measure the length of a rectangular table, but you can't find a tape measure. You do find, however, a sixty-centimeter measuring stick with no markings. Starting at one corner of the table, you keep placing the stick back-to-back along the table's longer side until you reach the other corner. The stick fits along the table's side three times, but it doesn't fit four times. This simple measurement tool lets you conclude that the table's side is between three times 60 cm, or 180 cm, and four times 60 cm, or 240 cm. Can you find a better estimate? You look into your junk drawer again, where you find a shorter, fifteen-centimeter stick, also unmarked. You repeat the exercise, and this time the stick fits thirteen but not fourteen times, which lets you estimate the length of the table's side as between thirteen times 15 cm, or 195 cm, and fourteen times 15 cm, or 210 cm. Had you found an even shorter stick, the mundane measurement process would drive you crazy—but you'd get closer and closer to the true length of 200 cm. In other words, for well-behaved shapes, such as a straight line, reducing the measuring stick's length gets us closer to the actual length of the shape. But this doesn't explain Richardson's findings.

Many country borders follow natural barriers such as rivers or mountain ranges, so they tend to be wiggly and way more complicated than a side of a rectangular table. Still, we can use the same measuring sticks to measure the border's length on a map. Pick a starting point along the border and place one end of the measuring stick there. Follow the border as closely as possible, counting how many times the measuring stick fits along it. When you reach the end, you'll multiply this number by the length of the measuring stick and convert it to kilometers or miles using the map's scale.

Now repeat the experiment with increasingly shorter measuring sticks. From our table experience, one might expect that the result will get increasingly closer to a single value—the true length of the border. Except it doesn't: the shorter the measuring stick, the longer the measured length, and this length increases without bound. This means that a natural border consisting of coastlines, rivers, and other wiggly lines has no single true length. We could even reasonably argue that its length is infinite.

All these measurements are as right as they are wrong. A shorter measuring stick lets you capture nooks and crannies that you omit with a longer one—and, consequently, increases the measured border length. For some lines, the results vary more than for others. For example, the length of the Norwegian coast, which is full of spectacular fjords, can double or even triple when we increase the scale. On the other hand, relatively smooth borders, such as the long straight beaches near Los Angeles, don't lose much of their length with decreased scales, so the discrepancy will be smaller. Grab your phone and zoom in and out on Google Maps to see that. We call this phenomenon of measurements' dependence on the measuring stick's length the coastline paradox.

We take measurements for granted—when we find a number in a reputable source, why would we question it? The more careful will look for measurement error, the inevitable part of any measurement process, which appears because neither humans nor the tools we use are perfect. Usually, the true figure will be close to the number cited, and the question is only up to how many digits. Richardson discovered that this isn't true for border lengths. Quite the opposite, as in this case more detailed measuring tools don't get us closer to a constant number, as we'd expect, but simply keep growing. The science of measurement had been established centuries before, and if we can't trust measurements, can we trust anything? After all, numbers underlie every scientific discovery, so doubting them could make us question the whole of science. Even Richardson didn't seem to realize how world-shaking his discovery was and didn't publish his results. So nobody noticed—almost.

Noisy Cauliflowers

From Benoit B. Mandelbrot's résumé, it's hard to infer that he would become one of the most important twentieth-century mathematicians. He earned his tenure only after turning seventy-five, which made him the oldest professor to receive tenure at Yale, an appointment that was an unusual end to a truly unusual career.

Born to a Lithuanian Jewish family in my hometown, Warsaw, Mandelbrot lived in Poland's capital for only the first eleven years of his life. In times of rising Nazism in Germany, Poland was a hostile environment for Jews, so Benoit and his family emigrated to France three years before the Second World War started. Luckily for young Benoit, his uncle Szalom Mandelbrojt, a mathematics professor already living in this unfamiliar country, took care of his education. Despite being mostly self-taught, young Mandelbrot passed the entry exams to the prestigious École Normale in Paris. However, he left the university after one day, unhappy with the faculty's contempt for visual thinking in mathematics, eventually graduating from an equally prestigious French school, the École Polytechnique.

He went on to gain a master's degree in aeronautics from the California Institute of Technology before returning to Paris to complete a doctorate in mathematics. For the rest of his career, he moved constantly between France and the United States, taking various research positions (and getting married in the meantime). Finally, he settled at IBM in New York, where he became interested in a special type of pattern that comes up in fields as diverse as abstract mathematics, biology, and economics.

At IBM, Mandelbrot was tasked with eliminating the unwanted disruptions—called the noise—from computer data transmissions over phone lines. The company thought that these errors were scattered at random, but Mandelbrot noticed that they appeared in clusters. When plotted, these clusters revealed an interesting property. A picture representing all the noise within one day was almost indistinguishable from a plot of errors from one week or even one month. This made the noise that bothered IBM a bit like . . . a cauliflower.

Imagine a whole, fresh cauliflower, and carefully detach one big branch—it looks just like the whole vegetable, doesn't it? From this branch, detach a smaller one, and again, you get almost a copy of the vegetable you bought from the market. You can keep going, although at some point the tiny branches will resemble flour rather than the initial cauliflower. In his 2010 TED talk, Mandelbrot described a cauliflower as very complicated and very simple at the same time. Smaller pieces of this vegetable are similar to the whole—not identical, but they look alike. Mandelbrot noticed the same property in the seemingly random noise he studied at IBM.

Mandelbrot thought back to weird geometric shapes that fascinated him as a young mathematician. His teachers used to say that "they were just esoteric constructions of no possible significance," but he couldn't shake the feeling that they might be the key to solving IBM's problems. To create one of these "esoteric constructions" called the Cantor set, one takes a line segment and removes its middle third, then removes the middle third of each remaining part, then repeats the process, again and again. The resulting shape resembled the noise at IBM that came in bursts of different magnitudes, each containing clusters of clear signals. The ratio of noise to signal remained constant, which explained the uncanny resemblance to the self-similar cauliflower. This was bad news for the company: the noise wasn't generated by guys with screwdrivers messing around with the network, as IBM's engineers tended to explain the problem. Mandelbrot understood that a "guy with a screwdriver" wouldn't create such a consistent pattern of disruptions. Instead, the self-similarity indicated a structural problem rather than a series of local interferences. This meant that the previous attempts to remove the noise by increasing the signal strength were counterproductive, and IBM was better off accepting some level of noise and focusing on developing systems that would catch the errors. Thanks to Mandelbrot's appreciation of visual thinking, "esoteric constructions" turned out to be useful.

Richardson Uncovered

Mandelbrot started noticing self-similar patterns in the most unexpected places. One day, he was clearing out his desk when he glanced at the appendix of the *General Systems Year Book* of 1961. Probably on the verge of throwing out this rather obscure publication, what he saw intrigued him enough to take a deeper look—which is how Mandelbrot was introduced to the old and forgotten work of Lewis Fry Richardson, published almost a decade after his death.

We'll never know if Mandelbrot cared the tiniest bit about Richardson's original research on the causes of wars. Instead, he focused on the nonintuitive and confusing discrepancies in reported border lengths. A result completely ignored by the scientific community was finally recognized for what it was—evidence that nature doesn't care about the standard, school-taught geometry.

For years, Mandelbrot had studied intriguing, self-similar structures such as transmission noise, but fellow scientists didn't take his research seriously. Whenever he suggested that some process in economics or physics, for example, required nonstandard mathematical tools, he usually got a response along the lines of, "When all the other fields of science can be tackled by proven mathematical methods from familiar textbooks, why should my field necessitate newfangled techniques, for which the only references are dusty tomes written in French, or even in Polish, or incomprehensible modern monographs?" In the coastline paradox, Mandelbrot saw a concrete and visual example of the complicated notion of self-similarity he'd struggled to convince people of for a long time.

In 1967, Mandelbrot followed up on Richardson's results, this time in a journal as well known as it gets: *Science.* Because every scientist worth her or his salt regularly glances at *Science,* Richardson's revolutionary insights finally saw daylight. While the title of Mandelbrot's paper wasn't too imaginative (although we've already seen worse), "How Long Is the Coast of Britain?" piqued the interest of fellow scientists. Because if *Science* decides to publish a paper on a seemingly trivial subject, there must be more to it. And there was. The subtitle "Statistical Self-Similarity and Fractional Dimension" revealed what this paper was about. Mandelbrot, taking Richardson's work as an example, introduced the concept of self-similarity and fractality (which is pretty much a fancy name for the wiggliness of coastlines and borders) to the wider world.

How Long Is the Coast of Britain?

There are multiple ways to define dimensions. In everyday speak, a dimension of space indicates the number of coordinates needed to specify a point within it. Imagine an ant walking along a line segment. To describe her position with respect to the beginning of the segment, the ant would need one number: how far she walked from the beginning. The same ant would need two numbers to describe her position in reference to a chosen corner of a sheet of paper—how far she walked up and to the right, for example. That's why a line has one dimension and a flat square two.

But this definition applies only to well-behaved objects such as smooth lines and surfaces. Let's take our ant to Dover, the coastal English town famous for its stunning cliffs. After a big little feast of fish and chips, the ant starts walking clockwise along the British coast, with a tiny pedometer on one of her wrists. At some point, our ant's best friend, who happens to be an elephant, calls her up because he wants to join her along the way. The ant shares with him the reading on the pedometer, so that the elephant knows how many miles from Dover she currently is. Being an English ant, she decides to wait for him over a pint in a local pub.

A few hours later, a disgruntled elephant calls her again. "Hey, I followed the coast for 300 miles, as you asked me to, but the pub is nowhere to be found," he complains.* Unfortunately, the elephant has walked way past the ant, who had followed every tiny bay and inlet of the coast, which were missed by the elephant's large steps. These differences quickly add up, so to get to the pub the ant walked a much longer distance than her big friend.

The same situation wouldn't happen on a perfectly circular island. With no extra wiggles to trace, the ant's route wouldn't be much longer than the elephant's. But the coast of Britain, unlike the coast of a circular island, isn't one-dimensional.

In his seminal *Science* paper, Mandelbrot suggested a different way of looking at "dimensions" and made an appeal to stop using the notion of length for such nonordinary curves altogether. This idea came from Richardson, but—as we know—the impact of his findings was completely ignored.

* I've yet to find a place in the United Kingdom where a pub is "nowhere to be found."

Measuring Curves

To measure a circle, which is devoid of straight lines, you can't use a straight measuring stick. What you can do is inscribe a triangle with sides of equal lengths into this circle, measure these lengths, and sum them up, which will give you the first approximation of the circle's length. But you can do better—inscribe a square. And then a pentagon. And hexagon. And heptagon—you get the idea. You'll notice that the results are getting bigger and bigger as you inscribe polygons with more and more sides. But the result doesn't increase forever; instead, it gets closer and closer to the true length of the circle (expressed by 2π times the radius).* This method works because a circle is an ordinary curve, with nice properties. Repeat a similar procedure with the coast of Britain, though, and the increase of approximations doesn't stop but grows to infinity. What's the difference between a circle and the coast drawn on a map?

Let's build Kochland, an imaginary island with a coast similar to the coast of Britain (minus the cliffs and views). We start from a triangle with all sides of length one. Now, we divide each side into three parts, draw equilateral triangles with the middle part as its base pointing outward, and remove the base. We repeat the procedure for each side of the new island. And again. And again, until we're too tired or we run out of ink (Figure 3.1 shows the first four iterations of the procedure). The wiggly border of Kochland resembles the coast of Britain. When we zoom into any part of Kochland's coast, we see . . . Kochland's coast. This means that this curve is self-similar (like the cauliflower). We call such self-similar, never-ending patterns *fractals*. Kochland's coast is in fact one of the most famous fractals, first described by Swedish mathematician Helge von Koch and known as the Koch snowflake or the Koch curve.

Since we can easily encircle Kochland, it must have a finite area. What about its perimeter? Each iteration of the procedure quadruples the number of sides—but the length of each side is one-third of the previous side's length. In each iteration, we multiply the perimeter of the island by $4/3$. For example, after four iterations Kochland's perimeter is $4/3 \times 4/3 \times 4/3 \times 4/3$ times larger than the original perimeter. The perimeter of Kochland grows indefinitely—it's infinite! So, we have managed to

*Approximating curves and surfaces by smaller and smaller line segments or polygons is a powerful technique of calculus.

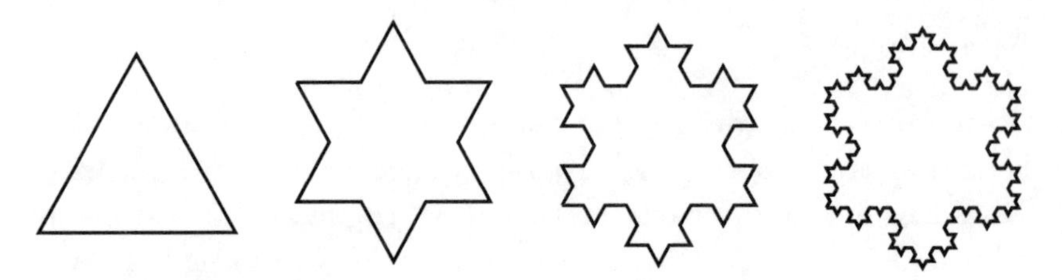

Figure 3.1 Creating Kochland, the first four iterations.

create an island with a finite area and infinite perimeter. And this is the source of the ant and elephant's problem. The ant was walking along the same coast as the elephant, just zoomed in. In the case of a circular island, zooming in doesn't increase the length of the trip, but with the coast of Britain, similarly to Kochland, the more one zooms in, the more one sees. The shorter the steps, the longer the coast.

Richardson observed a similar property when he looked at lengths of land borders. He measured different geographical curves with measuring sticks of decreasing lengths (which is more or less equivalent to increasing the map scale) and plotted the results against the measuring stick length. To fit all these vastly different measurements in one picture, he used a trick well known to astronomers classifying cosmic objects or geophysicists measuring magnitudes of earthquakes: he modified the plot's axes. On a usual, linear axis, a one-unit increase corresponds to adding one unit, for example, ten kilometers. Instead, on a logarithmic axis, a one-unit increase corresponds to *multiplying* by one unit, for example, by ten. In other words, logarithmic axes don't show the measured values but their magnitude, which is roughly the number of digits in a value.

Additionally, Richardson repeated the experiment with a simple circle. He noticed that the logarithmic plots corresponding to real-world curves resembled straight lines with various slopes, but the plot corresponding to a circle didn't (see Figure 3.2). The circle's length slightly increased with decreasing measuring sticks, but it quickly reached its true length. On the other hand, the lengths of geographical curves didn't stop increasing, and the wigglier curves increased faster. This observation made the standard formula we use to compute lengths—the length of the measuring stick multiplied by the number of times the measuring stick fits along the curve—all but meaningless. Instead, Richardson proposed a formula for the "length" of a curve that depended not only on the measuring stick but also on a special number D assigned to each curve.

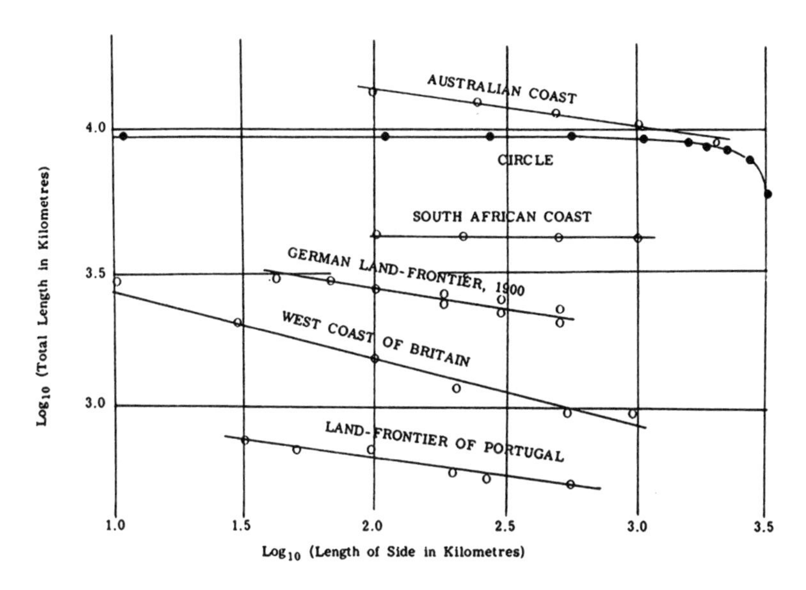

Figure 3.2 Data plotted by Richardson.

Richardson didn't think much of this unitless number. Later, however, Mandelbrot figured it could be treated as a special type of dimension that characterizes the curve's wiggliness, a bit like how density assigned to each substance describes how much mass fits in a unit of volume. The wiggliest curve measured by Richardson—the west coast of Britain—had D equal to 1.25. On the other hand, the coast looking the smoothest—the coast of South Africa—had D equal to 1.02. The truly smooth curve—the circle—had D equal to one, as we'd expect from a proper, ordinary one-dimensional curve. So, smooth curves are one-dimensional (with D equal to one), while wiggly curves have dimensions bigger than one.

Some curves are even two-dimensional, as they can fill the space. This is a big deal! If we drew these special curves with an infinitely thin pen on a two-dimensional page, we'd end up with a completely covered sheet of paper. Well, we would if we never stopped drawing, but still, the idea of a two-dimensional line is exciting. Google engineers share my enthusiasm.

Maps, including those online, show the world in two dimensions. The problem is that computers decode everything as a one-dimensional string of numbers, so to render a two-dimensional user-friendly image from these data, we need an efficient method of translating one dimension into two. Luckily, we have space-filling curves.

Google's solution is essentially to put the Earth in a box whose sides are covered by a grid of tiny squares. Then they draw a line passing through every cell of this grid, creating an approximation of a space-filling curve.* Finally, they project this curve onto our spherical Earth. Here's the trick: because the curve covers the whole box, the projected line goes through every projected square exactly once. This means that every place on the Earth is uniquely represented by a single number. Neat, isn't it? The king of Britain lives in 4979483113283, tourists climb Mount Kilimanjaro in 1664833531727, and Christ the Redeemer looks over Rio de Janeiro from 41204668869.

If lines can be one, one and a quarter, or even two-dimensional, what *is* a dimension?

Weird Dimensions

We intuitively know how to interpret whole-number dimensions: we understand zero-dimensional points, one-dimensional lines, two-dimensional squares, and three-dimensional cubes. Our intuition fails us the moment we encounter dimensions expressed as fractions. But their meaning is closer to the "normal" dimension than you think—it all has to do with scale.

Take any city plan, for example of Manhattan, and measure the length of a straight-line segment of your choice, such as Park Avenue between 23rd and 34th Street. Now find a plan of the same city but on one-third of the scale—this time, the same segment will be a third of the original length. So, for a one-dimensional object, one-third of the scale results in one-third of the length. Now find a rectangular area such as Central Park and measure it by multiplying its side lengths. What happens to this area when we make the scale three times smaller? It becomes $3 \times 3 = 9$ times smaller. Similarly, if we scale down a three-dimensional building, let's say Lincoln Center, its volume will go down $3 \times 3 \times 3 = 27$ times. We used the words "length," "area," and "volume" to describe the same concept: measure. We can see how they relate to dimension *D:* an object of dimension *D,* scaled down three times, has a measure 3^D times smaller.

*Drawing a true space-filling curve wouldn't be possible since even the most powerful computer cannot draw an infinitely long line. The smaller the squares of the grid, the better the approximation.

Let's repeat the same exercise for our imaginary Kochland. If you inspect its perimeter, you'll notice that each part of it is composed of four copies of itself but scaled down three times. This means that if we present Kochland on one-third of the original scale, the length of its perimeter will go down four times. In other words, Kochland's perimeter must have dimension D such that $3^D = 4$, which is about 1.26. It's neither a one-dimensional line nor a two-dimensional plane, but something in between, which fits nicely with Richardson's observations. The smoothest line he found—the coast of South Africa—has a dimension close to one, almost like a circle. On the other hand, the wiggly west coast of Britain fills the space to a greater extent than a line and has a dimension closer to two.

For each curve, Richardson found a number D such that the curve's length increases 2^D times if the scale doubles, 3^D times if the scale triples, and so on. He didn't, however, provide any physical interpretation of this number. It was Mandelbrot who recognized that D can be interpreted as a dimension and coined the term *fractional dimension.* While similar concepts had been used to describe theoretical curves and shapes, it was the first time someone connected the abstract notion of dimension with real-world phenomena, from coastlines to computer noise.

A "Trojan Horse Manoeuver"

Mandelbrot's revolutionary idea didn't fly with the mathematics community. He was ostracized, and it's hard to blame the mathematicians for their skepticism. We all have a deeply ingrained idea of dimension: we live in a three-dimensional world full of one-dimensional curves and two-dimensional surfaces. But 1.26-dimensional "lines"? This idea makes us feel uncomfortable. What do they look like? How do they behave? Do we even need such abstract objects?

This is exactly what even the most open-minded mathematicians thought not such a long time ago. In the annotations to his *Science* paper, Mandelbrot complained that "mention of a fractal dimension in a paper or a talk led all referees and editors to their pencils, and some audiences to audible signs of disapproval. Practitioners accused me of hiding behind formulas that were purposefully incomprehensible." Only the accidental discovery of Richardson's appendix helped Mandelbrot to change the general response. If borders and coasts—curves we know and understand—have these weird fractal dimensions, maybe the concept of a fractal dimension isn't as abstract as it seems? By disguising a deep mathematical theory behind a geographical curiosity—a "Trojan horse manoeuver," as he called it himself—Mandelbrot brought fractal dimensions to a wider audience.

Given the usual unenthusiastic approach to his work, the prompt acceptance of his paper submission to *Science* came as a surprise. The referee behind the positive feedback was the esteemed Polish mathematician, Hugo Steinhaus. Years before Mandelbrot, Steinhaus had already noticed the coastline paradox when measuring the length of the largest Polish river, the Vistula. In his 1954 publication, Steinhaus noticed that "the left bank of the Vistula, when measured with increased precision, would furnish lengths ten, a hundred or even a thousand times as great as the length read off the school map." No wonder that when Steinhaus saw his own insights, also completely ignored by the mathematical community, in Mandelbrot's paper, he promptly recommended that the paper should be published. He must have seen it as a chance for the coastline paradox, known from ancient times and noticed by multiple prominent mathematicians, to make it into mainstream research.

While we don't know who first noticed this measurement problem, apparently ancient Greek sailors struggled to establish which of the two (today Italian) islands, Sardinia or Sicily, is larger. Sailing around Sardinia took longer, but Sicilian fields were known to have bigger areas. It didn't make sense, but you have probably already guessed what was going on; a glance at the map shows that both islands have rugged coastlines. So, the crew of a large ship circumnavigating Sicily could easily report a shorter perimeter than a captain of a smaller boat able to stay close to Sardinia's shore. The big ship would sail along a smooth, almost circular path, while the smaller boat would follow the island's bays and inlets, increasing the reported length.

Mandelbrot wondered why this omnipresent measurement problem had been consistently ignored, especially given that Steinhaus, a world-class mathematician, had studied it in his paper. He refused to blame it on "the fact that no one reads Polish scientific journals—even those written in an international language" (a conclusion which would leave me personally offended). In Mandelbrot's opinion, Steinhaus himself hadn't considered his insights worthy of publication; they had found their way onto the printing press only because of the pressure on scientists to publish—and publish a lot. This might be the only case where the problematic "publish or perish" approach, so common in academia, has resulted in something positive.

Maybe Mandelbrot wasn't the first to notice the need for fractal dimensions, but he was the first to understand its importance. An excellent science communicator, he managed not only to popularize fractals among fellow mathematicians but also to bring them to the wider world. Today, we find fractals everywhere, in medicine and in meteorology, in art and in pop culture.

Fractals, Fractals Everywhere

I was too young to attend the premiere of *Toy Story* but old enough for *Toy Story 2.* I still remember the cinema trip, one of the first in my life. My dad had graciously volunteered to endure two hours of talking toys, and there we were, sitting in the darkness in front of the largest screen I had ever seen, with a bucket of greasy popcorn that he had bought me after I'd promised I wouldn't fuss at dinner. A big Disney fan (to this day), I had already seen plenty of cartoons, but this Pixar production was different. Sheriff Woody looked so realistic, as did Andy's room (including his adorable globe) and the Pizza Planet delivery truck in which the adventure started. Little did I know that soon the technology behind this outstanding animation would get the Sci-Tech Academy Award of Merit—given alongside other Oscars—and that this technology was fractals.

In 1980, a newly minted computer scientist named Loren Carpenter was helping Boeing visualize their planes during flights. His task was to show the machine from different angles against a backdrop of mountains because, in his own words, "every Boeing publicity photo in existence has a mountain behind it." But there was a problem. Any hiker would agree that mountains are fractals—at least, any hiker who happens to have read this book. Every big peak has its smaller peaks, and each smaller peak has even smaller peaks, and so on, down to little rocks. To visualize a mountain range, the poor computer would have to store information about all these shapes, which would have been impossible at a time when the most powerful machines had less memory than today's smartphones.

Here's where Carpenter comes in. He stumbled upon the book *The Fractal Geometry of Nature,* in which Mandelbrot had described how fractals work and included pictures of fractally generated landscapes. This gave Carpenter a brilliant idea: instead of storing the information about all parts of a mountain range, the machine would need only the fractal pattern to visualize the whole landscape. By definition, a fractal is a repetition of itself, on different scales. The math was there, but a computer algorithm—that is, a set of precise instructions to generate the landscape—was missing. Carpenter rose to the challenge, and the results stunned even him. After hearing that the world-famous Lucasfilm was about to open a computer graphics department, he was determined to get in. He realized how ambitious a goal it was and knew that the only way to get a job offer was to create a film. And create he did.

Carpenter worked relentlessly on a two-minute animated movie, *Vol Libre,* which became the first film to use the fractal algorithm. The film takes us over a mountain range that looks surprisingly realistic given the limited resources of a Boeing engineer in 1980. That July, Carpenter presented his invention, together with the short film, at a prestigious computer graphics conference. The head of Lucasfilm's computer design team was so impressed that he offered Carpenter the job on the spot. "I hired him immediately because he had an efficient algorithm for rendering landscapes and other complex phenomena and because he also had a sense of mission, like the rest of us, to make movies and tell stories with technology," Alvy Ray Smith commented on his decision.

Together with the small team of rising stars hired by Lucasfilm, Carpenter went on to develop his technology. And that's how RenderMan, the first software able to visualize realistic landscapes, was born. The new tool became the key to the success of *Star Trek II: The Wrath of Khan,* as it was now capable of creating not only mountain ranges but whole realistic planets. By using fractals, the team animated the characteristic scenery to look plausible from all possible viewpoints. RenderMan took off, and the team itself was renamed Pixar.

If visual effects in a movie look realistic, you can safely assume that the secret behind them is fractal-based software, RenderMan or similar. From Forrest Gump's incredible ping-pong game to realistic dinosaurs in *Jurassic Park* and icy landscapes in *Frozen,* Mandelbrot's insights changed cinematography. RenderMan became the first software to receive an Oscar—or four Oscars, to be precise. One of the statues went to the three men behind the fractal-based software, Loren Carpenter and his Pixar colleagues Ed Catmull and Rob Cook, "for significant advancements to the field of motion picture rendering as exemplified in Pixar's RenderMan." In 2015, Pixar generously released the award-winning RenderMan for free noncommercial use.

The applications of fractals aren't limited to the entertainment industry, of course. Today, fractals save lives, as doctors hunt for divergences from a healthy, fractal heartbeat or distinguish normal tissue from a nonfractal tumor. Meanwhile, engineers build fractal antennas for the best signal strength to area ratio, mimicking the optimal, fractal shapes designed by nature: from our lungs to trees, to river deltas, to DNA folds. As Mandelbrot famously said, "clouds are not spheres, mountains are not cones, coastlines are not circles, and bark is not smooth, nor does lightning travel in a straight line." Once you see it, you won't be able to unsee it. Unfortunately, the failure to notice fractals can have dire and long-lasting consequences.

A Long Dispute

When I think of Alaska, I see a large American state attached to the northwest of the Canadian province of Yukon. If you asked me what its border with Canada looks like, I'd respond that it's simply a straight line. Well, almost. Southeast Alaska, or the Alaska Panhandle, is a strip of coastline and offshore islands so narrow that it's easy to miss on a small-scale map. It looks as if a child has been coloring in Alaska on a map and accidentally left a few spots of paint on the Canadian coast—and, at least according to Canadians, that's not so far from the truth.

For the European used to maps with the Americas presented on the left-hand side, it might seem that there's nothing to the west of Alaska beyond the narrow Bering Strait that separates Alaska from the Chukchi Peninsula, Asia's easternmost peninsula. At the narrowest point, only about fifty-five miles separate the United States and Russian territories. Given this geographical situation, it shouldn't be a surprise that it wasn't America but Russia who first colonized the coast and islands of today's southeast Alaska, no doubt attracted more by the waters abundant in tasty fish and sea otters used for expensive fur than by the breathtaking views. In the late eighteenth and early nineteenth centuries, Russian settlers claimed the area as part of the Russian Empire. Brits and Americans soon followed suit, moving in with no regard for the native Łingít and Haida communities.

Given so much interest in the abundant waters of the Alaska Panhandle, it became time to formalize the region's boundaries. In 1825, in the beautiful city of Saint Petersburg—known for its magnificent canals and often called the Venice of the North—Russians and Brits signed the Convention Concerning the Limits of Their Respective Possessions on the Northwest Coast of America and the Navigation of the Pacific Ocean (a name that says it all). According to the treaty, 54°40′ N latitude defined the southern border of the Alaska Panhandle.

But what about the eastern boundary? Nobody cared. Both nations wanted access to rich Alaskan waters and had no interest in the barren—so they thought—interior. But, just in case, they agreed that "the line of coast which is to belong to Russia . . . shall be formed by a line parallel to the winding of the coast, and which shall never exceed the distance of ten marine leagues [thirty-five miles] therefrom." Good luck drawing a border on the map based on this vague description!

Three decades later and half a world away, Russians and Brits—the latter with France, the Ottoman Empire, and Sardinia on their side—were fighting for the Crimean Peninsula. Weakened by the lost war, the Russians didn't feel up for another possible conflict with the Brits that was likely to happen in Alaska, so they sold their territory to the United States. After the Alaska Purchase, as it is known today, Canadian officials asked their southern neighbors to survey the Panhandle, so that the border could be settled once and for all. This sounded like too much effort to US officials—maybe they thought they'd have to leave their comfortable Washington offices and venture out into the cold and unwelcoming mountains of southeast Alaska. They refused to spend taxpayers' money on setting a boundary in some faraway, "uninhabited" (because native people didn't matter, it seems) land. It was a big mistake.

At the end of the nineteenth century, the Yukon territory, today's Canadian province of Yukon, became the real-world El Dorado. In particular, Klondike—a region close to the US–Canada border—turned out to be rich in gold. This gold had to be transported, so Canadians would have appreciated some control over access to the coast.

The Treaty of Saint Petersburg placed the border thirty-five miles east of the coast. But what is the coast? The coastline paradox makes the fjords of the Alaska Panhandle particularly hard to deal with. The treaty didn't specify the scale of the map on which the border should be drawn, and the multitude of offshore islands didn't help. Washington, to push the boundary as far inland as possible, argued that the coast is where the mainland meets the ocean; Ottawa, to reach the opposite goal, wanted to define it as the western boundary of the islands. After they failed to resolve the conflict on their own, in 1903 an international tribunal had to intervene. Three Americans, two Canadians, and one Englishman convened in London to pore over maps provided by both sides: thirty-four by the Brits and forty-six by the Americans. In the end, the Canadians were shocked when Britain, a representative of the colonial government, supported the United States. In protest, Canadian delegates refused to sign the agreement, but this didn't stop the tribunal from publishing a new atlas with a clearly marked boundary.

At first glance, the new Alaskan border is a compromise between the two claims, as it's drawn between the two suggested boundaries. A closer look, however, allows us to see that the United States gave up mostly mountainous areas, while Canada was forced to cede the harbor they had fought for.

Why did this dispute happen in the first place? The 1825 treaty specified the part of the border between the United States and Canada between the latitude 56 degrees north and longitude 141 degrees west. According to this document, the boundary should follow "the summit of the mountains situated parallel to the coast," unless the summit is further than ten marine leagues (about thirty-five miles) from the ocean, in which case the boundary should be parallel to the coastline instead, always within ten marine leagues from the coast. I hope that at this point you're asking this question: what map scale should be used?

The boundary could follow all the deep and narrow inlets and bays of the Alaskan coast, or it could cut across these interruptions. Even measured in a straight line, these fjords can stretch out for miles, so different interpretations of the treaty's vague statement would lead to considerably different boundaries. And this is just one of the imprecise statements that were supposed to define the important—as it later turned out to be—border. In total, the tribunal formulated seven questions that needed answering, including where the border began, how it should be measured, and which mountains the treaty was talking about.*

The coastline paradox caused a major boundary dispute that, as some scholars claim, led to anti-British feelings among Canadians, which

*For this chapter, the most relevant is question five. Its lengthy statement can be summarized: Should the boundary run around the heads of the bays, ports, inlets, havens, and waters of the ocean as opposed to cutting through them? The answer was affirmative. Of course, since the treaty didn't specify the map scale to be used, the problem wasn't fully solved.

contributed to their campaign for more autonomy from the Crown. Geopolitics was the last place where I expected to find math—and I was wrong.

Measure and Rule

It is late evening in Warsaw when Ryan Stoa, a law professor, connects from his new home in Louisiana to join me on a Zoom call. After reading his paper on the legal implications of the coastline paradox, I knew I had to talk to him. The Alaskan boundary dispute, it became clear, is just the tip of the iceberg. When I ask Stoa about his research, he says that his "big picture surprise . . . was how many times the length of a coastline really matters."

Oceans abound in natural resources, from fish to oil, which humans have exploited since time immemorial. Even back in the seventeenth century, a Dutch humanist and lawyer, Hugo Grotius,* had insisted that oceans should belong to all the people, but that coastal nations should own the most adjacent waters. This idea was later specified as the "cannon-shot rule," according to which each state with sea access can claim waters as far as a cannonball shot from land can reach—in practice, about three nautical miles.† As the scale of ocean exploitation grew, it became increasingly important to form a proper international agreement indicating who these resources belong to.

The first relevant document was signed in 1958 at the United Nations Geneva Conference on the Law of the Sea. It specified the exact lines from which national waters should be established, called "baselines." In general, it stated that coastal states should measure the distance from the coastline itself. However, a country with particularly wiggly coastlines or a multitude of offshore islands can opt to connect coastal points with straight lines, essentially including parts of the sea in its territory. Anticipating that states might abuse this allowance, the document limited the straight-line rule to situations where the coastal points connected by a straight line lie at most twenty-four nautical miles apart, and if a semicircle drawn between them encircles less of the ocean than the bay itself. Intuitively, this means that the coastal state can claim that a bay constitutes its internal waters, as opposed to the ocean, if the bay significantly cuts into its land territory. By

* His real name was Hugo de Groot, but as we already know, Latin was cooler.

† Nautical miles are slightly (1.1508 times) longer than the miles some nations use in everyday life.

establishing this additional rule, the conference recognized the coastline paradox without naming it.

The agreement from Geneva sparked a decades-long discussion that resulted in the United Nations Convention on the Law of the Sea (UNCLOS), as of today ratified by 168 parties. The hundreds of articles specified who the oceans belong to and from what baselines coastal nations can measure zones of varying degrees of the coastal country's jurisdiction: the territorial sea,* the contiguous zone,† and the exclusive economic zone,‡ all illustrated in Figure 3.3. Again, these territories should be established from the coastline, or straight lines when it's impractical. Unfortunately, UNCLOS failed to specify these special cases precisely enough, leading to coastal nations abusing the permissive law. The vast majority of coastal states applied straight-line baselines, regardless of the coast's actual wiggliness. Why wouldn't they? A straight line will always encompass more of the ocean than a line parallel to the shore, giving the state the areas of bays and inlets that would be lost otherwise. One can't argue with simple geometry.

In 1951, a decades-long dispute between the United Kingdom and Norway was taken to the International Court of Justice. The Fisheries Case, as it's known today, stems from ambiguous sea laws and the coastline paradox. Norway is famous for its beautiful fjords along its long coastline that exceeds 1,500 miles even when measured in a straight line. The magnificent fjords not only attract tourists (who leave a lot of money in the area, as Norway is one of the most expensive countries in the world) but also supply plenty of fish. British fishermen had sniffed out the opportunity for a rich catch centuries earlier, infuriating the king of Denmark and Norway so much that at the beginning of the seventeenth century, they were forced to give up fishing in Norwegian waters. But three hundred years later, at the beginning of the twentieth century, the Brits came back to get their

*The territorial sea is a belt of coastal waters up to twelve nautical miles from the baseline that is part of the coastal state's territory. In this area, the country has full jurisdiction, although it must guarantee passage rights to other states.

†A coastal state might establish a contiguous zone that extends beyond the edge of territorial waters, up to twenty-four nautical miles from the baseline. While the contiguous zone doesn't count toward the country's territory, it helps the coastal state with law enforcement, for example by preventing its criminals from escaping the territorial sea.

‡The exclusive economic zone is an area of the sea that extends the territorial waters of a coastal state up to 200 nautical miles from the baseline. In this zone, the state has exclusive rights and responsibilities to explore, exploit, and conserve natural resources.

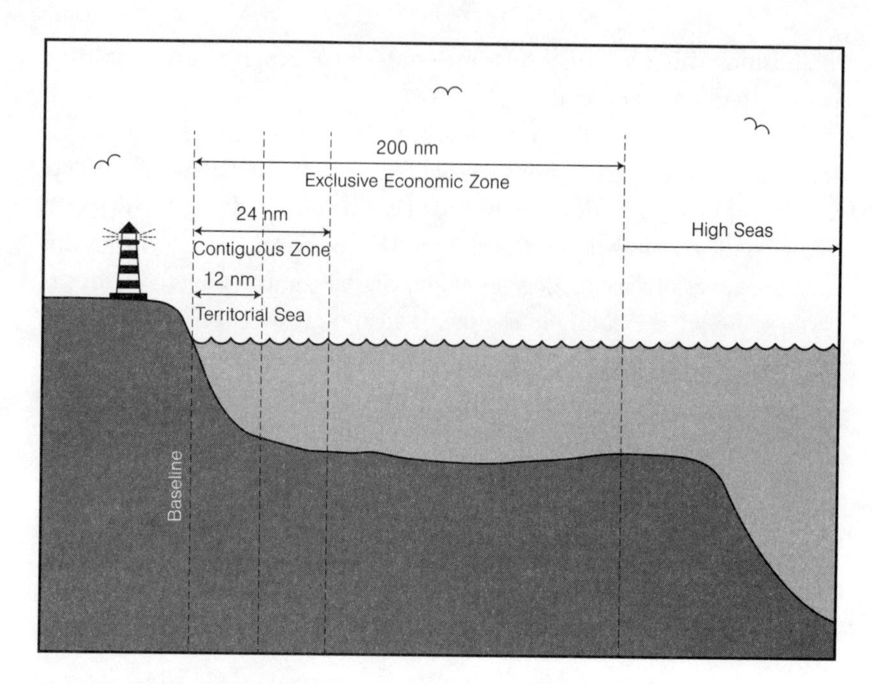

Figure 3.3 Maritime zones established by UNCLOS.

share of fish, this time not on wooden boats but wholesale fishing ships, which started a fishing war between the two countries.

Norway insisted on their exclusive right to fish within twenty-four nautical miles from the straight baseline, while the United Kingdom wanted this boundary to follow the narrow inlets cutting deep into the land. This is the equivalent of the UK establishing the Norwegian territory on a large-scale map, with Norway preferring to use a small-scale map. This time, the court stood for Norway, encouraging other coastal nations to follow suit and fight for their territory.

The coastline paradox is not just a mathematical curiosity. It has a tangible impact on international law. As Stoa tells me, "Many times, policymakers or negotiators come across the coastline paradox and understand the problem without actually naming it as the coastline paradox or understanding that it is the coastline paradox." In other words, people intuitively recognize the ambiguities of the simple word "coastline" without being familiar with fractals and dimensions. Mathematics impacts our life whether we're aware of it or not.

The legal implications of the coastline paradox aren't all happening on an international level. In the United States, for example, a significant portion of the funding for coastal management is proportional to the length of the state's coastline. This means that the money a state receives depends more on the length of the measuring stick used to measure the coastline than on the state's population size. So, it's not unusual for small Maine to get almost as much funding as large California. The same goes for estimating the cost of flood insurance, as the endangered area depends on the coastline length. And if you happen to own a large waterfront property, you might want to check how the local government measures its boundary for tax purposes—the shorter the measuring stick, the longer the coastline, and the larger your tax.

Although the coastline paradox impacts who the money and the power belong to, few policymakers or even scholars devote their time to thinking about ways of fixing this problem. To avoid international boundary disputes, underfunding of coastal areas, and unfair under- or over-taxation, we first need to recognize the issue and then agree on a uniform length measure. Because for now, as Stoa says, we're "comparing apples to oranges without people really knowing they're comparing apples to oranges," which isn't ideal. The length matters more than we think.

The Power of the Scale

It might seem that the difference between maps of various scales comes down to how big the objects appear on a piece of paper. The consequences of this choice, however, go way beyond whether we need glasses to find our destination. The larger the scale, the more details appear on the map. For example, the lengths of the same line—be it a border, a river, or a coastline—measured on maps of different scales will differ widely. This observation started a new field in mathematics and changed the way we think about dimensions.

We're surrounded by objects of weird dimensions, both in nature and technology. Fractals gave us new ways to diagnose diseases, create realistic animations, and build efficient antennas. Despite their omnipresence, these creatures remain largely invisible, and that invisibility has tangible consequences. The lack of awareness of the coastline paradox leads to lengthy international disputes and misplaced funding of coastal areas, which impacts our day-to-day life. We ignore fractals at our peril.

4

DISTANCED

How to Navigate

Among Londoners' fears, Tube strikes rank high. Every day, millions of commuters rely on the dense public transportation system to get to work, drop kids off at school, and return home just in time for dinner. So, when in February 2014—long before working from home was an acceptable and manageable practice for many—the UK's capital faced a two-day-long strike of underground network workers, nobody expected any positive outcomes.

On February 5, almost two-thirds of Tube stations were closed, but commuters still needed to get places, which forced them to come up with alternative routes. Surprisingly, after the strike was over, some preferred to stick to these new commutes, which took less time than their regular journeys. It turned out that quite a few Londoners had never considered whether their usual routes were optimal in the first place. We're creatures of habit and tend to experiment only when forced to. But why is it so difficult to pick the fastest Tube line?

The Tube Map

Both Londoners and tourists rely on the famous London Tube map to plan their journeys. Everyone who has ever visited the UK's capital is familiar with the neat network of colorful straight lines, connecting at the major stations and separating again. But the Tube map as we know it didn't always look this way.

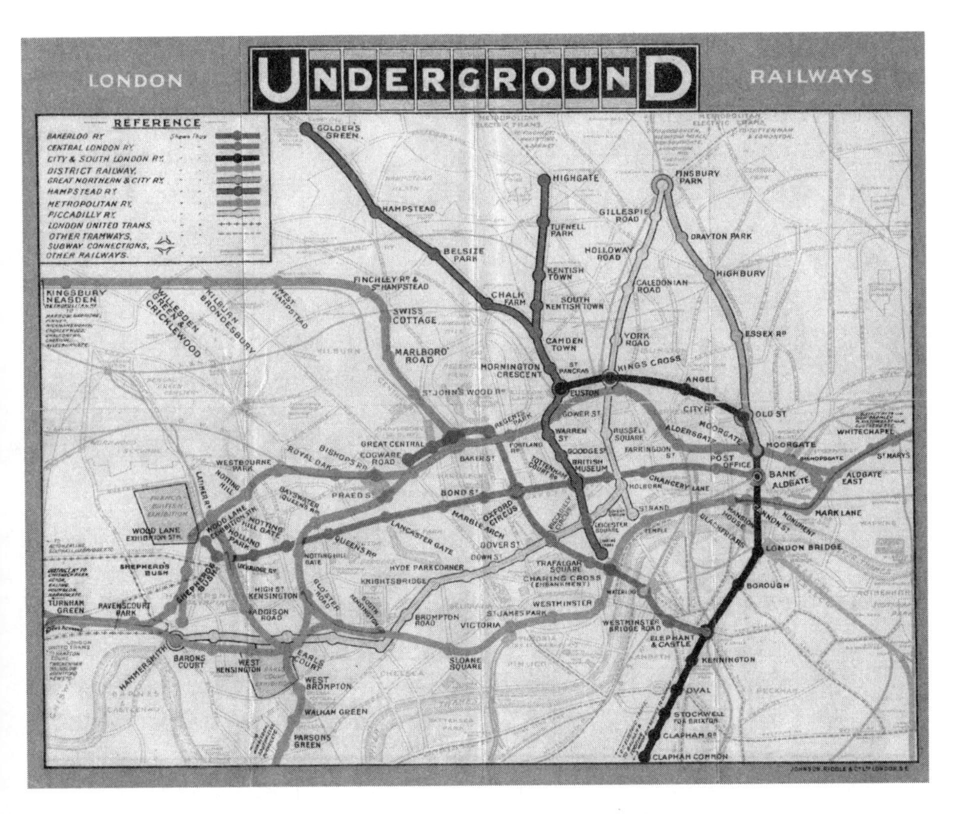

Figure 4.1 Pocket Underground map issued by Underground Electric Railways of London in 1908.

In the second half of the nineteenth century, London's growing public transport system, consisting mostly of horse buses, demanded increasingly detailed maps. By 1908, London Underground expanded so much that it required a separate map.

The map overlaid underground railways on top of the city plan. This natural approach worked well enough for a few decades, but the map of the rapidly developing Underground system was getting increasingly crowded and difficult to read. At some point, the Tube map became so complicated that it discouraged Londoners from using this fast way of transportation for fear of getting lost—until Henry Beck saved the day.

In the mid-1920s, young Henry Beck, today affectionately referred to as Harry, started working as an engineering draftsman for the London Underground Signals Office. His job was to draw diagrams of the Tube's electrical systems, which involved a combination of artistic and engineering skills. As a contractor, Beck had some quiet periods, during which he didn't waste his time. A commuter himself, fed up with the confusing Tube map, Beck sketched a new version.

Beck's map did away with geographical details. "If you're going underground," he said, "why do you need to bother about geography? It's not so important. Connections are the thing." It's hard to disagree—when we step on a Tube train, we don't care what's above us, as long as it takes us to the next station. Beck removed all unnecessary details, such as streets, parks, and buildings, leaving only the Thames. Then, he straightened up the wiggly Tube lines so that they all ran vertically, horizontally, or diagonally, at forty-five-degree angles. To make the commuter's choice even simpler, he assigned a separate color to each of the untangled lines.

The Underground's publicity department considered Beck's design too unconventional and for some time kept the geographically accurate maps. In 1933, however, Beck came back with an improved version, and this time his idea was printed on 500 trial pocket maps. It was such a hit with Londoners that new pocket editions and posters soon followed. Beck's map changed the public image of the Underground system from messy and complicated to clear and simple to navigate. By reducing the distances between remote stations, he encouraged Londoners to move away from the crowded city center to the suburbs, which didn't seem so distant anymore. And, maybe most importantly, Beck's design became a poster child of a new branch of mathematics.

Play-Doh Mathematics

We've seen that some map projections distort shapes, distances, and areas, and how this can be confusing. Sometimes, however, such distortions are helpful and welcome.

In an area of mathematics called topology*—from the Ancient Greek *topos* ("place, locality")—exact shapes and distances are unnecessary details that obfuscate interesting properties. Topologists consider two objects the same if one can be obtained from the other without any cutting, tearing, or gluing. So, when a topologist's child smashes a Play-Doh ball with their fist, their parent won't notice a difference. Only when the kid creates two

*Not to be confused with topography, topology is a branch of geography that studies the physical features of land surfaces.

smaller balls out of the bigger one will the topologist see two different shapes.

I like to think about topology as a study of connections. When we stretch, bend, or squeeze an object, making sure not to tear or glue it, it seems that we lose all its properties. But one thing persists—such deformations, called homeomorphisms (from the Greek *homoios* meaning "resembling, equal" and *morphē* meaning "shape"), preserve connections between points. If two points are close before, they'll still be relatively close after the homeomorphism, which is exactly the idea behind Beck's map.*

Beck took London Underground lines and bent, stretched, and squeezed them as he pleased, but he kept the properties crucial for commuters: the connections and the order. If two stations lay on the same line, in a particular order, they also appeared so on Beck's map. If all that commuters cared about was where they were and where they needed to arrive, Beck's topological London Tube map gave them exactly what they wanted.

As Beck's map distorts properties such as relative distances between stations and the directions (compare the real-world Thames with its depiction on the Tube map), how do we know that it preserves the connections and the stations' order? One could carefully check these two properties for every single line and every single Tube station—but why waste time if we have topology?

The standard school mathematics curriculum doesn't stress enough that mathematics is all about abstraction. Mathematicians study specific objects but, usually, they aim to cluster them together and study them all at the same time. This saves huge amounts of time and effort because they must prove properties only once, for the whole class of objects, instead of studying each object individually. Even more importantly, this *guarantees* that every object within a class has a particular property, even the objects that haven't been defined yet. Often, when mathematicians want to prove that an important object behaves in a particular way, they prove instead that it belongs to a class already known to have this property. We reason similarly in everyday life: since all fruits have seeds, and an apple is a fruit, then an apple must have seeds.

*Sometimes a homeomorphism requires moving to a higher dimension to avoid cutting, tearing, or gluing. For example, imagine two circles touching at one point, which together resemble the shape of the number 8. It is homeomorphic with two circles touching at one point, where the smaller one is inside the larger one. But to get one figure from the other, we need to "lift" one circle, so we cannot do it in a two-dimensional space.

For a more mathematical example, imagine you want to know if the sum of two whole numbers (1, 17, 100,000, etc.) is odd or even. You can figure it out easily if the numbers are small: $1+3$ is even, and $2+7$ is odd. The larger the numbers, the more cumbersome the addition becomes, but luckily, you don't have to add them at all. It turns out that the sum of two odd or two even numbers must be even, while a sum of an odd and an even number must be odd, regardless of the exact values. This property of odd and even numbers spares us from unnecessary additions.

Similarly, we can be sure that Beck preserved connections and the order of the stations because the transformation he applied was a homeomorphism—he only bent and stretched the real-world lines, without any forbidden moves, such as tearing or glueing. Since topologists have already proven that all homeomorphisms preserve connections and the order of points, we can be sure that these properties hold for all London Underground lines and stations.

Helpful or Confusing?

Beck's map was such a success that today Transport for London and other transport authorities in cities all over the world regularly release updated topological maps of local transportation systems. Thanks to the topological Tube map, we don't have to consider complicated London geography when planning how to get to our destination—but maybe we should.

The same map, too, contributes to passengers repeatedly choosing suboptimal routes, unless forced to change their habits by events such as a Tube strike. Shaun Larcom and colleagues studied how individuals moved before, during, and after the 2014 Tube strike. By comparing true distances between stations to distances represented on the Tube map, they identified parts of London with smaller and bigger distortions. It turned out that commuters who lived or worked in considerably misrepresented parts of London were more likely to stick to the newly discovered route after the strike. This indicates that the map convinced some commuters to choose a suboptimal route and only the forced experimentation during the strike made them question the supposed efficiency of that route.

The Tube map is a great tool if we only want to get to our destination, but it becomes useless if we care about the speed. For example, imagine that you just arrived at Paddington Station, world-famous thanks to the adorable bear, and want to get to Bond Street Station. You have two options:

you can either pass underneath Sherlock Holmes's residence on Baker Street or picturesque houses next to Notting Hill Gate. According to one study, about one-third of all passengers chose the latter, even though it's a significantly slower route. It doesn't, however, look slower on the Tube map, which creates a neat, symmetric rectangle out of a more complicated layout.*

The 2014 Tube strike gave commuters a chance to discover that their "optimal" route wasn't as optimal as they thought. While in the end, only about one in twenty commuters chose not to return to their previous routine, the time gains of this small group were so large that they surpassed the total time wasted by the rest of Londoners due to strike disruptions.

Use with Caution

It seems that topological transportation maps prevent confusion and confuse us at the same time. On the one hand, they make it clear where we are, and which line will take us to the destination. On the other hand, they don't give us the information needed to choose the fastest route. So, should we love them or hate them? We should certainly be aware of the mathematical properties of the map we're looking at.

When I moved to London, I took the Tube map for granted. It's so iconic that one can find it not only at Tube stations and inside trains but also in souvenir shops, on T-shirts popular among tourists, in art galleries,† and even as a scar above Albus Dumbledore's left knee. Viewers of BBC Two's *The Culture Show* voted Beck's Tube map as the second favorite British design of the twentieth century, preferring only the famous shape of Concorde. You can't blame me, then, for never questioning the suspiciously straight lines of the underground network in such a geographically complex city. I realized that something was wrong when I decided to walk quickly from White City to East Acton in west London, assuming that these two stations are as close to each other as Leicester Square to Covent Garden, so it shouldn't have taken me more than three minutes. Almost half an hour and a few sorry-I'm-late calls later, I finally reached my destination. I thought

* The study predated the opening of the Elizabeth line, which connects these two stations directly.

† For example, check out Simon Patterson's lithograph *The Great Bear* in Tate Modern, where station names were replaced by names of stars—as the title suggests—although the stars are human: actors, politicians, scientists, and so on.

back to all the times I had bumped into another passenger on a train taking a sharp turn on a supposedly straight route, and I finally saw the smart use of topology in the map I had considered myself familiar with.

Topological maps are excellent tools when we use them for the purpose they were created to serve: to pick the Tube line that will take us to our destination. They're especially useful in cities with complex layouts, such as London—not so much in more regular, gridded New York or San Francisco. On a recent trip to San Francisco, I appreciated the (mostly) geographically accurate train map when, exhausted after my long flight, I debated leaving the train at an earlier station to grab some food before checking in at the hotel. Without the backup of ridiculously expensive data roaming, I wouldn't have dared such an experiment had the map been topological—this could have ended with the equivalent of the White City to East Acton hike, plus a large suitcase and a huge sleep deficit. But this map worked only because, as opposed to London's Tube, the San Francisco train network approximately follows a rectangular grid, so its geographical map doesn't look as messy as London's would.

Topological maps aren't a new idea. One of the few surviving maps from colonial times in North America created by Native Americans, not the colonizers, is an excellent example. The early eighteenth-century "Map of the several nations of Indians to the Northwest of South Carolina" is today known as the Catawba Deerskin Map, since it was drawn on deerskin, most likely by Indians of the Catawba Nation. It was presented to the colonial governor of South Carolina, Francis Nicholson, who then ordered two copies and took them to London, thus preserving the map for future generations, even though the original didn't survive.

The map represented the Indian nations that lived between Charlestown (today Charleston, South Carolina) and the colony of Virginia. The squares and straight lines showed European settlements and the circles depicted Indian nations. Similarly to Beck's Tube map, the Catawba Deerskin Map ignored the distances and directions and instead emphasized the connections between different Indian nations and the English settlements. This map wasn't meant for navigation, but for trade and diplomacy, for which the relationships and connections mattered more than geography.

Ironically, we lost so many cartographic treasures made by Native Americans because Europeans considered their maps primitive due to their geographic inaccuracy, only to recreate their structure in modern trans-

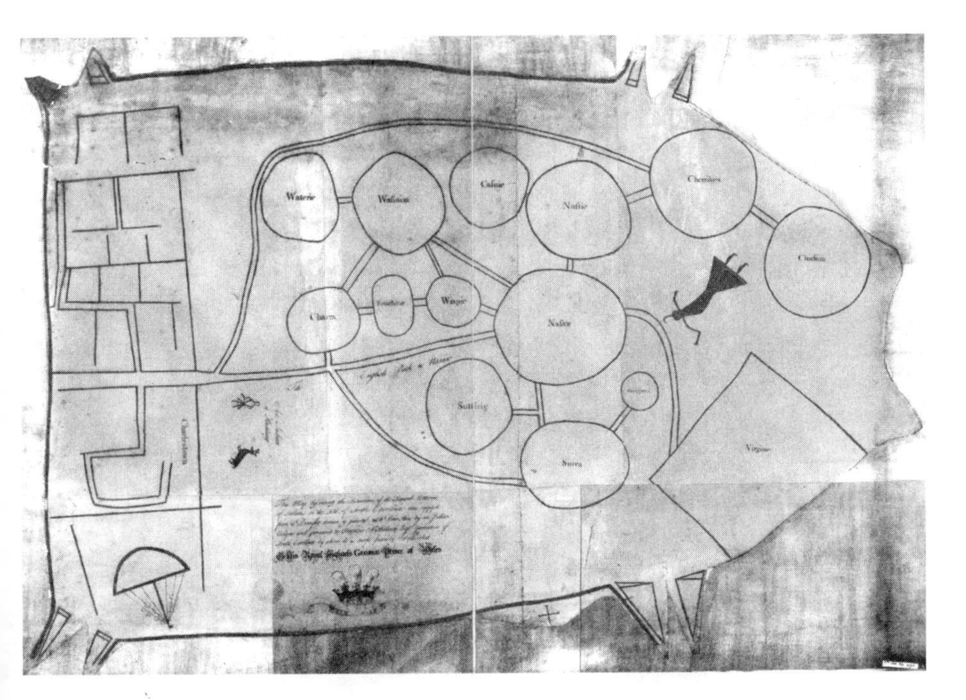

Figure 4.2 The Catawba Deerskin Map.

portation maps. It seems that Native Americans, for centuries considered inferior by the Europeans, figured out the power of topological maps long before their colonizers.

The Catawba Deerskin Map is by no means the oldest example of a topological map. Placed on UNESCO's Memory of the World Register, "Tabula Peutingeriana," also known as the Peutinger Map, represents the impressive road network of the Roman Empire. While the surviving map, today safely stored behind closed doors at the Austrian National Library in Vienna, was created in the thirteenth century by a Dominican monk, historians believe it to be a copy based on the fourth- or fifth-century original.

To fit the vast road network of the Roman Empire, the map is about one foot high and almost twenty-two feet long, divided into eleven parchment scrolls for preservation.* This massive map contains at least 555 cities

*In 1911, an extra sheet was added to include the British Isles, the Iberian Peninsula, and parts of North Africa missing from the surviving original.

and 3,500 other place names, including rivers, mountains, forests, seas and, most importantly, over sixty thousand miles of road.

Because of its unusual format, the east–west distances were drawn at a much larger scale than north–south distances. For example, the map stretched the tall-but-narrow boot-shaped Italian peninsula in the west–east direction, and somehow the island of Rhodes migrated next to Tel Aviv. In addition, not all directions were accurate. If the south–north direction of the Nile had been preserved, for example, it wouldn't have fit on the map, so it was "redirected" to flow eastward. But the Peutinger Map never pretended to be geographically accurate. It looked like a modern transit map because it had a similar goal—to represent the connections between the cities in the Roman Empire. In other words, it was topology that mattered, not topography.

For all their distortions, topological maps are arguably the best tool for navigating transportation networks. When we get on the Tube, we need to make sure that the line goes through our destination station, but we trust the driver to figure out the exact way to get us there. When we need to walk, however, the distances start to matter to us.

Legible London

In the early 2000s, each of the thirty-two boroughs of London had a separate signage system, each incompatible with the others, which made navigating the city by foot a nightmare. As a result, neither locals nor tourists were keen to walk and risk getting lost; instead, they tended to choose less environmentally friendly modes of transportation. And those brave enough to walk would often attempt to find their way using the Tube map,* which—as we already know—is less than ideal for this purpose. In 2004, the then Mayor of London, Ken Livingstone, set out to put an end to this madness and turn London into a pedestrian-friendly city by 2015—and the idea of Legible London was born.

After numerous studies and public consultations, Transport for London installed trial signs in nineteen central locations. The experiment was so successful that the signs spread all over the city; today it's hard to spend any

*According to Transport for London's study, over one-quarter of pedestrians tried to navigate using the Tube map. Another popular navigation tool was an A–Z guide, a small book format designed for driving rather than walking.

amount of time in the capital without noticing them. Legible London consists of over 1,500 signs placed in carefully chosen positions within the city. While many signs appear inside Tube stations, on bus shelters, and at bike rental stations, the system focuses on the simplest mode of transportation: walking.

Most recognizable are local maps with circles of appropriate radii, centered at the viewer's position and marked "You are here." These indicate how far the pedestrian can get within five or fifteen minutes—or at least that's the intention. The problem with these recognizable circles is that they assume the pedestrian's ability to walk through walls or fly over buildings, rivers, and other urban obstacles. The circles are drawn assuming the average walking speed* and a straight-line distance, which is a rather bold assumption in a city with such a complicated layout. One of the most extreme cases is the sign next to the Ravenscourt Park Tube station in west London. A neat fifteen-minute-walk circle crosses the Thames as if the pedestrian didn't first have to take a detour to cross Hammersmith Bridge before heading to the destination on the other side of the river. This makes the true walking time significantly different from the one depicted on the map. Indeed, some of the areas captured in Transport for London's fifteen-minute-walk circle take as long as half an hour to reach (see Figure 4.3). Are distances in the city at all meaningful? It depends on how we define them. To test that, let's visit another iconic city, this time on the other side of the Atlantic.

A Mathematician in New York

Walking along the orderly streets of Manhattan, it's hard to believe that only two centuries ago most of its area was covered by farms, meadows, ponds, and marshes. Back then, the vast majority of nearly 100,000 New Yorkers lived on the southern tip of the island. As the city expanded, new streets were built one by one, without any plan or vision, but at some point, this approach became unsustainable. So, in 1807 the city hired three commissioners to come up with a more structured idea for Manhattan. Simeon De Witt, Gouverneur† Morris, and John Rutherfurd published the so-called Commissioners' Plan of 1811, in which they designed a rectangular

*This is a different problem—in fifteen minutes fast walkers will get much farther than slow walkers.

†Gouverneur was Morris's first name, derived from his mother's surname, not his position.

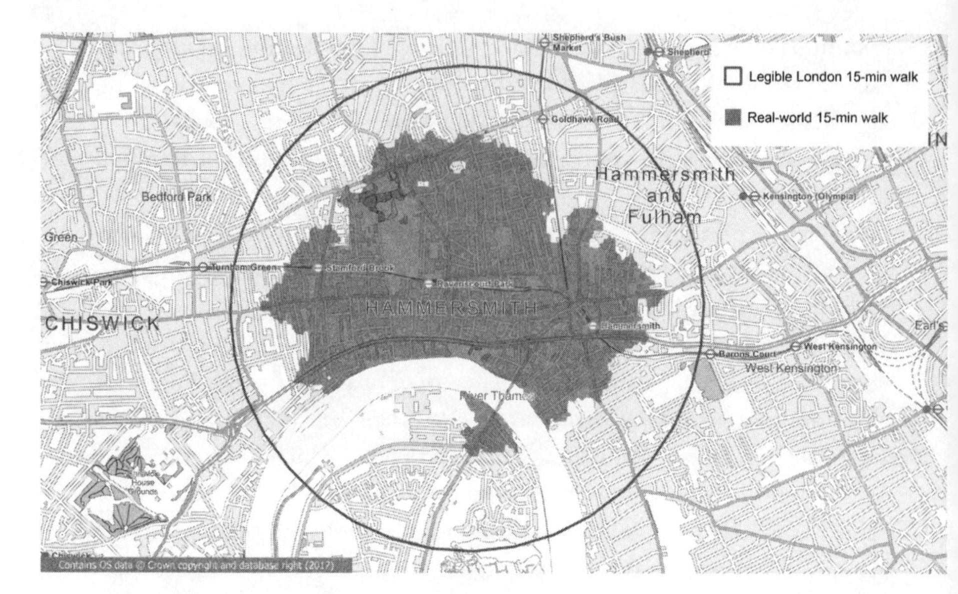

Figure 4.3 A fifteen-minute walk from Ravenscourt Park Tube station in west London: Legible London versus reality.

street grid stretching from Houston Street in the south to 155th Street (which now separates Harlem from Washington Heights) in the north. In other words, they presented the neat layout of avenues running southwest–northeast and streets running northwest–southeast that we use today.

Once the plan was approved, John Randel was put in charge of the gargantuan task of its implementation. Many residents of the island didn't welcome Randel and his team. Some feared that, with a single decision, the surveyors might force them to relocate, transforming their land into a new street. Some simply hated the idea of an ordered and repetitive grid. The antipathy toward the surveyors was so strong that, on more than one occasion, Randel was arrested for trespassing. Once, an angry resident attacked him and his team with cabbages and artichokes, forcing the state to offer them legal protection. The completion of the project took about sixty years, and the result became the beginning of the Manhattan we know now.

Today's Manhattan, with its world-famous museums, theatres, and skyscrapers, attracts millions of visitors each year. The most efficient—and, in my opinion, the most pleasant—way of sightseeing is to walk between the places of interest.

Imagine that, feeling cultured after watching an opera at Lincoln Center, you decide to peek at one of the most iconic structures in New York City: the Chrysler Building. Using a city plan, you find that the distance between these two attractions is 1.5 miles. Well, this would be true if Manhattan consisted mostly of meadows and farms, but today's dense architecture of Manhattan won't allow you to walk in a straight line. Instead, you'll need to follow the gridded streets and avenues, which will increase the distance by one-fourth, to about 1.9 miles.

When someone asks us about the distance between two places, in response we should ask what distance they have in mind. In dense, built-up areas, straight distances are of limited use. What we're usually more interested in is the walking or driving distance, which is almost always larger than the straight-line distance, unless you want to walk across an empty field.* While underestimating the distance between tourist attractions simply extends your evening stroll, a similar mistake could have dire consequences if you were, for example, a first responder rushing to an accident. And yet, many studies of ambulance or fire brigade travel times still involve the inappropriate straight-line distance.

Of course, the straight-line distance is still a useful way to express how far apart two places are. Mathematicians refer to it as the Euclidean distance, to commemorate the Greek mathematician Euclid of Alexandria, whose textbook *Elements* served students for about two thousand years. However, the Euclidean distance is only one of many possible ways we can measure closeness.

In mathematics, we express how close two objects are with functions called *metrics* or *distance functions.* Different metrics are appropriate in different contexts, but they all take two points and return a single number—the distance between them. For example, if we feed the Euclidean metric with Lincoln Center and the Chrysler Building, the function will return 1.5 miles, but other metrics might return different values. Does that mean that any mathematical function can measure distances? Not really—every metric must satisfy a few crucial properties.

Let's think for a moment about what we expect from a distance. First, if you ask how far Lincoln Center is from Lincoln Center, it's reasonable to

* Careful readers will point out that measuring a straight-line distance on a globe means measuring a distance along a great circle. However, because in this chapter we're considering tiny (in comparison to Earth's radius) distances within a city, we can ignore the Earth's curvature.

get zero, regardless of the metric. Conversely, if I tell you that I'm going to a place whose distance from Lincoln Center is zero, you'd consider me crazy if I was going anywhere else but to Lincoln Center itself. So, any distance from a point to the same point should be zero, and any metric should return zero *only* when both inputs are the same.

Second, we want the distance from Lincoln Center to the Chrysler Building to equal the distance from the Chrysler Building to Lincoln Center. This is true for the Euclidean distance but, in the real world, this property might not hold, because one-way roads might make the outward journey shorter or longer than the return. However, mathematicians like to simplify things, so metrics must return the same number no matter the order in which the two points are inputted.

Finally, taking deviations shouldn't make the route shorter. In other words, the most direct journey should give the shortest distance between two points. We call this last property of metrics the "triangle inequality" because of what it says in geometric terms: for any triangle, the sum of the lengths of any two sides (the sum of distances from the departure point to the mid-journey stop, and from the mid-journey stop to the destination) cannot be smaller than the length of the third side (the direct journey), as shown in Figure 4.4.

This gives us three properties that any distance function must have:

The distance from point P to P is zero, and the distance from P to point Q is zero only if P is the same as Q.
The distance from P to Q is the same as the distance from Q to P.
The sum of distances from P to R, and from R to Q, cannot be smaller than the distance from P to Q.

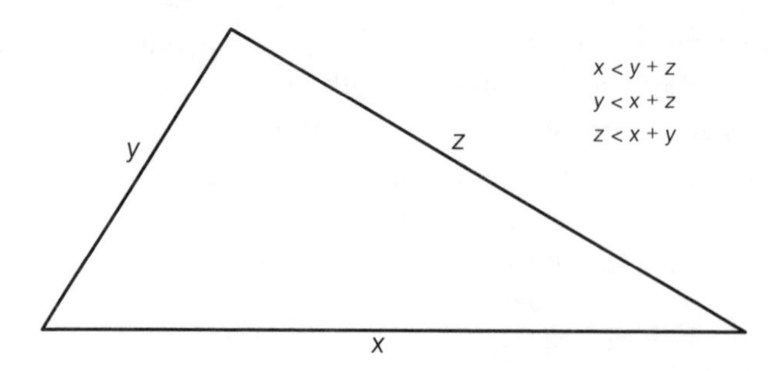

Figure 4.4 In a triangle, the sum of lengths of any two sides is greater than the length of the third side.

Many functions, useful in different circumstances, satisfy these three properties. One of the most popular metrics is particularly appropriate in our New York example: the Manhattan metric.

To use the Manhattan metric, we need to assume an idealized version of a gridded city—let's call it Gridhattan—that consists of one-by-one blocks determined by roads running north–south and east–west. While Manhattan doesn't fully satisfy this assumption—for example, blocks vary in size and Broadway runs diagonally—the resemblance is so striking that it's no wonder this metric bears such a name.*

Imagine that at midnight, jet-lagged and exhausted after a long journey, you arrive at Gridhattan Central, the city's main train station. You're not too excited about the prospect of dragging an oversized suitcase to the hotel. Your hosts proudly advertise being only ten blocks away from the train station, so you head south, counting blocks between you and your comfy bed. After eight blocks, your phone instructs you to turn right, and you can't wait to see your hotel in only two more blocks.

Alas, the hotel is nowhere to be found. Exhausted, you keep going, eventually reaching your hotel after a fourteen-block journey. Furious, you demand to speak to the manager. This false advertising is outrageous! The manager does her best to calm you down, takes the plan of Gridhattan and explains what has happened.

To reach the hotel, you have walked eight blocks south and six blocks west, which adds up to fourteen blocks, as Figure 4.5 shows. The manager points out that you didn't take the shortest route, which is the straight dotted line between the train station and the hotel. Then, she prompts you to compute this distance using the Pythagorean theorem, which isn't something you tend to do at 1 a.m. Reluctantly, you recall that in a right triangle, the square of the side opposite to the right angle (the dotted line) equals the sum of squares of the sides next to the right angle (the solid lines): $?^2 = 8^2 + 6^2$. After solving this equation, you realize that the distance between the station and the hotel indeed equals ten blocks!

So which distance is the correct one? Both, but in different metrics: the Euclidean metric (along the dotted line) and the Manhattan metric (along

*It's unclear who called this type of metric the Manhattan metric. According to Google Ngram Viewer, the first recorded use of this term was in *Proceedings of the 1962 Workshop on Computer Organization,* published in 1963. An alternative name is taxicab distance, also inspired by a real-world application.

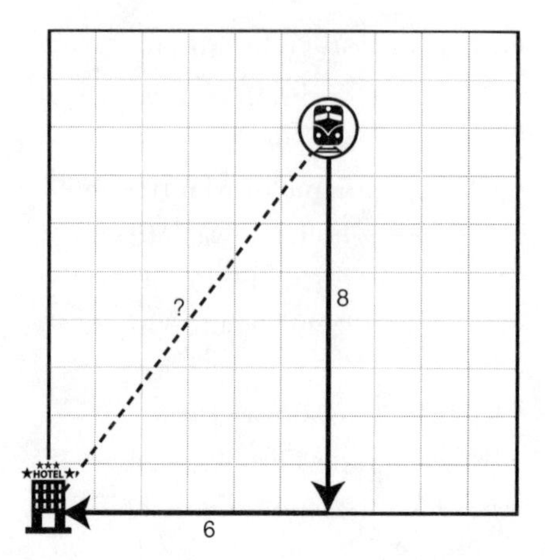

Figure 4.5 Distance from the train station to the hotel: Euclidean (*dotted line*) and Manhattan (*solid line*).

the solid lines). The manager is right that the distance along the shortest path equals ten blocks, so, strictly speaking, the hotel's advertising isn't wrong. However, guests can't fly over buildings, and the shortest way along the streets is fourteen blocks.

Abstract and Useful

Two people who are asked about the distance between the same two points on the same map might give two different but both correct answers. Also in mathematics, the notion of distance isn't unique. For a mathematician, distance is the output of any function defined on some set of objects (we'll see some examples shortly) that turns a pair of elements from this set into a number and satisfies the three conditions that we've already seen: it's zero if and only if the two points are the same, it's symmetric, and it satisfies the triangle inequality. A set together with a metric defined on this set is called a *metric space*.

Why do we bother defining a general metric space if we could separately analyze each of these functions—for example, the standard straight-line distance metric or the Manhattan metric? We can, of course, and we often do as we begin to study them but, as we've already seen, mathematicians like to generalize. The nice thing about having a whole category of objects that all have some properties—like the three properties we just listed—is that we can analyze them all at the same time. So, when we prove a theorem about metric spaces, we do it only once, without checking whether it holds for every single metric space (did I mention that we can define infinitely many metrics?). Quite convenient, isn't it? That's the reason why, in 1906, a French mathematician by the name of Maurice René Fréchet introduced the concept of metric spaces, although it was a few years later that a German mathematician, Felix Hausdorff, gave them their current name. But applications of distances go way beyond quantifying how far places are from each other.

In 1947, Richard Hamming was working at Bell Labs (today Nokia Bell Labs, due to its purchase by the Finnish telecommunications company Nokia), the hub of telecommunications research, with headquarters in New Jersey. On Fridays, as he was leaving the lab for the weekend, he would insert punch cards—pieces of paper with little holes that, back in the day, inputted data into machines—into his computer to start a new set of computations. Alas, on Mondays, as he arrived at work, ready to analyze the results, he'd often find that an error had crept into the computations, so he wasn't able to report anything useful to his colleagues. At this point, most researchers would take a deep breath and start the process again from scratch, but Hamming had had enough. "Damn it, if the machine can detect an error, why can't it locate the position of the error and correct it?" he asked, and with that question, he revolutionized information theory.

Hamming set out to invent a way for the computer to detect *and* correct an error, which would allow it to continue with the computations. Computers translate anything we input into long sequences of zeros and ones called bits. This means that as long as the machine can detect an error, it can also correct it—change a zero into one, or one into zero.

In 1950, Hamming published a seminal paper, "Error Detecting and Error Correcting Codes," in which he described his method of error detection and error correction. But before we dig into his idea, let's talk about my experience calling a bank when I first moved to the United Kingdom.

"Could you spell your last name for me?" a high-pitched voice with a thick northern accent asked.

"Sure, it's ar o double-u—"

"You mean Romeo Oscar Whiskey?" the voice interrupted.

"No, my name is Rowińska, so it's ar o double-u i en ess kay a," I insisted.

"OK, could you confirm that your name is Romeo Oscar Whiskey India November Sierra Kilo Alpha?"

At this point, I gave up and decided that the person on the other side of the line was making fun of me. Later, I learned that they were using the NATO phonetic alphabet, in which each of the twenty-six letters of the English alphabet is assigned a unique code word, to ensure the correct spelling of my name. To minimize the possibility of any miscommunication, the code words were chosen carefully, so that no two sound in any way similar. Indeed, it's hard to confuse *Romeo* with *Oscar* or *India* with *November*, even when you're speaking to a clueless customer with a strong Polish accent. This way, any word that can be written in the English alphabet can be encoded with the NATO phonetic alphabet with a minimal chance of miscommunication.

Richard Hamming's idea was similar, but he was dealing with computers, not humans, so his "words" consisted of zeros and ones instead of letters. For example, a computer might send a message "001100." If there's some noise, the message might turn into "00110**1**," but the machine has no way of telling whether any distortion occurred. If we repeat each bit, however, the computer will detect an error, as long as errors don't happen too often. With this method, "001100" will become "00 00 11 11 00 00." Now, if one bit gets distorted—for example, the computer receives "00 00 11 11 00 0**1**"—then the machine can tell that something went wrong since both bits within each pair must be identical.

Repeating each bit ensures error detection (under a reasonable assumption that two errors in a pair of bits are unlikely), but not error correction. In our example, the computer would figure out that the last part of the message has been distorted, but it wouldn't be able to tell if the last bit in the original message was supposed to be a zero or a one. A way around this would be to triple each bit instead of doubling it, so "001100" would be sent as "000 000 111 111 000 000." In this case, if the message got distorted to "000 000 111 111 000 0**1**0," the computer would not only detect but also fix the error—since the last triple has more zeros than ones, it's more likely that the original bit was a zero, not a one. By tripling each bit, the computer can both detect and correct errors.

So, problem solved? Tripling the digits is only useful if we don't care about the speed of message transmission. While repeating each bit three times doesn't seem like a big deal, this method triples the time spent on each communication. If we think of all the technologies that rely on error detection and correction, we realize that this is a problem. Imagine that your always-too-slow internet connection gets *three times* slower while you're trying to stream your favorite TV show, and you'll understand why Hamming decided to come up with a more efficient solution.

Hamming's idea resembles the NATO phonetic alphabet. While the phonetic alphabet defines a code word for each letter of the English alphabet, Hamming created a binary code word consisting of a fixed number of zeros and ones for each possible four-bit block of a message sent by a machine. Like in the NATO alphabet, no two of Hamming's code words were similar. But while we can all agree that it's hard to confuse "Romeo" with "Oscar," how can we tell if "0101011" and "0101100" are similar for a computer or not? To make this concept precise, Hamming defined a new type of mathematical distance: the Hamming distance.

Given two-bit sequences of equal length, the Hamming metric simply counts the number of bits that differ between the two sequences. For example, "0010110" and "0001010" are the same in the first, second, sixth, and seventh place, and they differ in the third, fourth, and fifth place. Since they differ in three places, the Hamming distance between "0010110" and "0001010" equals three.

The Hamming distance allows us to measure how similar two sequences are for a machine. Hamming code* transforms four-digit blocks of zeros and ones—such as "0101" or "1111"—into longer seven-digit sequences that are all at least three units apart, in terms of the Hamming distance. This way, the machine can detect that something went wrong if one or two bits in the original message were to get distorted and correct the code if one error appears.

For example, the code word for "1010" is "1011010." If the computer receives "**0**011010," it knows that something went wrong since it cannot find "0011010" among the code words. Then, it looks for the code word closest in terms of the Hamming distance to the distorted message—"1011010"—and translates it back to the original "1010." This procedure relies on the fact that there's only one code word closest to the distorted message, which saves the computer from unnecessary dilemmas. This is also the principle behind the NATO phonetic alphabet. If for some reason you hear *Pravo,* you'll immediately figure out that something went wrong, as this word isn't among the code words. By looking at all the words, you'll notice that the closest codeword to *Pravo* is *Bravo,* so you'll understand that someone wanted to

*There are infinitely many Hamming codes, but since they all rely on the same idea, here I describe the original Hamming(7,4) code.

communicate the letter B. If things went awry, *Alpha* could become *Pravo* too, but it's far less likely. Assuming that transmission errors are relatively rare—as they luckily are—both the NATO phonetic alphabet and Hamming's idea let us identify and correct errors by choosing the code word closest to the received message.

Hamming's result was so important that Bell Labs prevented him from publishing until they were granted a patent. If you're wondering how on earth one can patent a mathematical formula, you're not alone. "I didn't believe that they could patent a bunch of mathematical formulas. I said they couldn't. They said, 'Watch us.' They were right," Hamming recalled. Because of this delay, the first person to publish Hamming's idea wasn't Hamming but a Swiss mathematician, physicist, and information theorist, Marcel J. E. Golay, who independently generalized Hamming's error-correcting codes. This resulted in a decades-long priority controversy. Given the impact of error-correcting codes and abstract distances on our lives—for example, they let us receive satellite pictures from Mars, identify genetically related organisms by comparing DNA strings, and use QR codes—the authorship was worth fighting for.

It might seem that by defining metric spaces, mathematicians complicated the intuitive, map-inspired notion of distance. However, as we'll see next, these abstract, theoretical distances are still simpler to understand than the distances we keep—quite literally—in our minds.

All in Your Head

So far in this chapter, we've seen two different types of maps that aid urban navigation. On the one hand, we have topological Tube maps that show which train to take but don't tell us anything about the distance to the destination. On the other hand, we have metric city plans that let us figure out precise walking routes and measure distances. However, we still haven't talked about one type of map, arguably the most important, without which we'd be helpless in any environment: the cognitive map.

In 2014, the Nobel Prize in Physiology or Medicine was awarded to London-based John O'Keefe and a Norwegian couple, May-Britt Moser and Edvard Moser, "for their discoveries of cells that constitute a positioning system in the brain." All three Laureates had spent much of their neuroscience careers observing rats moving freely in a room, with tiny contraptions implanted into their brains to measure the activity of cells. These experiments contributed to our understanding of the sophisticated navigation system found first in rodents and, subsequently, in humans.

In 1971, O'Keefe and his student Jonathan Dostrovsky discovered that when rats entered a particular place in the room, certain cells in the hippocampus—a seahorse-shaped structure deep in the brain—became active. Other cells of the same type activated when the rat moved to a different part of the room. Hugo Spiers, a cognitive neuroscientist at University College London, explains to me that these cells, also later discovered in humans and named *place cells,* send "you are here" signals to the rest of the brain. By figuring out which place cells are active at a given moment, we can determine the rat's position to an accuracy of five centimeters. Place cells, working together, build a cognitive map of the environment. As if this wasn't impressive enough—it's pretty cool that we walk around with GPS inside our heads!—place cells are not the only navigation tool in our brains.

Over three decades after the discovery of place cells, the Mosers and their students found a different type of cell in the entorhinal cortex, an area of the brain neighboring the hippocampus. While place cells were triggered when the rat entered a specific location in the room, each of the newly discovered cells became active in many locations. But these locations didn't seem to be random—instead, they formed a regular hexagonal (six-sided) pattern, which inspired the researchers to name their discovery *grid cells.* A grid cell fires when the rat is at any location in the grid. Unlike a place cell, a single grid cell doesn't provide enough information to locate the rodent. However, since different grid cells correspond to different, overlapping hexagonal grids, a combination of them uniquely determines a rat's—or, as it turned out later, a person's—location. In other words, grid cells provide us with inner coordinate systems.

The idea of a cognitive map appeared decades before the discovery of place cells. In the 1940s, an American psychologist, Edward C. Tolman, experimented by putting rats in a maze and teaching them a roundabout route to a tasty snack. When he was confident they had learned it, the researcher modified the maze so that the familiar path was blocked. After realizing that they had to change their approach, the rats didn't follow the closest available path but instead planned a shorter route. This inspired Tolman to suggest that rats' brains mapped the environment, creating a cognitive map. Thanks to the rapid development of tools such as fMRI,* Tolman's idea has been confirmed and generalized to other species, including humans. But one question remains: what does a cognitive map look like? Does it resemble a topological Tube map, or does it encode precise distances in some kind of a metric like a plan of Manhattan? Some scientists insist that it's one or the other, but it's likely the truth lies somewhere in between.

In 1978, O'Keefe and his colleague Lynn Nadel postulated that cognitive maps are metric, which would mean that they preserve real-world distances and directions. They specified the metric as Euclidean, which I find rather surprising—intuitively, I'd expect maps formed by a brain to consider real-world rather than straight-line distances. The discovery of grid cells, which seem to correspond to the Euclidean coordinate system, supported O'Keefe and Nadel's idea.

Other researchers, however, have shown that people violate the properties of metrics we explored earlier. For example, external cues such as junctions, landmarks, and the number of turns we take impact our estimation of the distance and direction. We also repeatedly violate the symmetric property of a metric by judging the distance to the destination as different to the distance from the destination. Other researchers suggest that cognitive maps violate triangle inequality, as we're terrible at taking shortcuts. Also, we're so used to right angles (just take a look around you!) that we remember angles of various measures as ninety degrees. In particular, we aren't surprised when we return to the same spot after taking *five* ninety-degree turns, even though this would create an impossible pentagon. While all these inconsistencies have been observed for decades, neuroscientists can now test them by designing non-Euclidean virtual reality environments. It turns out that people have no problems navigating these impossible spaces and, what's even more puzzling, we often don't notice that something is off.

All these arguments provide compelling evidence that cognitive maps don't have a metric structure. Do they then resemble distance-free, topological Tube maps? That doesn't have to be true either. While our distance estimates are far from perfect, we do have *some* sense of distance. Otherwise, how would we explain our preference for shorter routes and the more or less successful search for shortcuts? Most of us have a rough sense of

* Functional magnetic resonance imaging (fMRI) detects changes in blood flow that occur in active regions of the brain, thus measuring brain activity.

distance and direction, even in the absence of external information. So, comparing cognitive maps to Tube maps doesn't seem to be accurate.

Proving or disproving the Euclidean hypothesis seems all but impossible. It wouldn't be reasonable to expect that a map created by our brain will denote perfect directions and distances, and the line between a distorted metric map and a topological map is blurry. An American psychologist, William H. Warren, offers an alternative image of the cognitive map, a structure between a distance- and direction-free topological Tube map and a metric city plan: a labeled graph that he calls a *cognitive graph.* In the next chapter, we'll talk about graphs in more detail; for now, we can think about them as points ("vertices") connected by lines ("edges"). According to Warren's hypothesis, at first glance the cognitive map looks like a Tube map, with landmarks connected by paths like stations connected by Tube lines. But this structure has additional, crucial elements: edge labels denoting approximate distances between landmarks, and vertex labels showing approximate angles between routes. As opposed to topological maps, labeled graphs would allow us to find shortcuts, but only approximately. Precise distance estimations would require a metric map.

A cognitive graph would also be easier to build than a metric map. Think about the last time you were exploring a new city without a map or a GPS. Most likely, you ventured out of your hotel and followed a street until you reached a landmark, let's say a cathedral. Then, maybe, you turned into a different street until you reached a local market. After this initial exploration, your brain formed a vague picture of the city: "The hotel is connected to the cathedral, and when I turn by ninety degrees from the cathedral and walk for about twice as long, I'll reach the market. So, I could probably reach the market faster by walking straight from the hotel in a slightly different direction." Notice that to form this brain map, you didn't need to know how the three landmarks—the hotel, the cathedral, and the market—fit into the whole city. In particular, you didn't need to know if the hotel is north or south of the cathedral, for example, which would have been necessary if you were to build a metric cognitive map. Such local exploration of new environments seems to work better for creating labeled graphs rather than global, geometric city plans. And yet, had you created a topological map, you wouldn't have been able to figure out the shortcut from the hotel to the market. This experience suggests that our internal navigation system indeed resembles more a labeled graph than any of the extremes: a topological map or a metric map.

Another possibility is that our brains are creating both maps and graphs, depending on the environment. Creating a metric map requires a reference frame, easily available in places with visible landmarks such as skyscrapers or mountains, or with straight routes. It seems that we're more likely to think in the Euclidean way in open spaces, where large landmarks are visible, or in gridded cities. On the other hand, in environments where we have to change direction a lot and where we often lose sight of major landmarks, we tend to navigate by planning the route in relation to local sights ("turn left after the train station, and then right behind the post office"). In other words, the cognitive map becomes more like a Tube map in cities with complicated street layouts, dense forests, and mazelike buildings. (From my experience, for some inexplicable reason, math departments tend to be examples of the latter.)

Another factor that might impact how we navigate is the size of the environment. It might be easier to generate metric maps of smaller spaces—especially if the whole area is visible at all times—than of large and complex environments. This could explain the popularity of the hypothesis of metric cognitive maps in the early days of spatial neuroscience when the most common subjects were rats and people navigating a single room. With advances in virtual reality technologies came the possibility of studying how we find our way in multicomponent spaces, such as multiple rooms connected by corridors. In such places, we tend to focus on *how* to get from one part to another rather than *how far* these two parts are, which can be nicely illustrated with Tube-map-style graphs.

Finally, there are individual and cultural differences in our navigation styles. I, for one, tend to think of cities and buildings in terms of graphs. I might think to myself, "The wide road will take me from the metro station to this nice square with cafes, while the narrow one will lead me to a dodgy area." But I have little sense of how all these landmarks fit into the whole city. In contrast, one of my friends confidently turns into a street she's never walked before because her well-developed cognitive map tells her where this road leads.

Researchers have also identified sex and gender differences in our navigational preferences—keep in mind that they correspond to an average woman and an average man rather than individuals.* In general, women

*Unfortunately, to the best of my knowledge, nonbinary people haven't been included in any of these studies.

prefer to follow familiar routes and orient themselves by landmarks, which suggests that we create cognitive graphs, and men are more likely to take shortcuts and head straight to the destination, which requires a metric cognitive map. These differences, however, are most likely cultural rather than biological. They're most pronounced in countries with large gender inequalities, and almost disappear when girls and boys experience a similar upbringing.

When it comes to the neuroscience of navigation, with new answers come new questions and controversies. While some researchers assume that animal and human brains create metric maps and some suggest that we prefer topology-based navigation, the truth likely lies somewhere in the middle.

Connections, Distances, Directions

I'm finishing this chapter on a Sunday afternoon, sitting over a slice of exceptionally tasty apple strudel and a cup of coffee in a quaint cafe in central Warsaw. Before leaving the apartment, I and my partner independently checked the route—and ended up with two different plans. He suggested taking tram 33 and changing onto tram 9 at the main train station, while I explained that we should walk south for about five kilometers, then turn east and walk for one more kilometer. Both ways would have taken us to our destination, but while he planned it in terms of connections, I was more interested in distances and directions; in other words, he used a topological map, and I looked up a metric city plan. (We ended up walking, in case you're wondering.)

Topological and metric maps help us see our cities from two different perspectives. Although human-made, both navigation tools have biological roots, as it seems that our brains use variations of these two strategies depending on the environment and individual preferences. Over centuries, mathematicians have built theories that explain and generalize these two types of maps. This way, we can apply the insights from problems as intuitive as navigating Manhattan to more abstract questions such as error-correcting codes. Topology illustrates the power of generalization: it lets us solve many diverse problems at once, saving us precious time and resources.

5

CONNECTED

How to Simplify a Map

An almost completely blind father of thirteen,* Swiss-born and Russian Empire-based Leonhard Euler wasn't the most likely candidate to become one of the most prolific mathematicians in history. And yet, in the seventy-six years of his life, he published over 800 works that added up to about 30,000 pages in total, in topics ranging from astronomy to calculus to geometry. One of his great achievements was turning a simple map-inspired problem into a study of connections, which changed the way we think about mathematics.

Before Euler, mathematics had been considered a science of numbers and measurements. He realized that precise values sometimes obfuscate the important parts of the problem. Instead, it might be the connections between locations, people, or even abstract concepts that matter, as we've seen with the Tube map. To understand how this revolutionary idea enables us, for example, to receive packages in time, plan weddings, and color maps, we'll go back to eighteenth-century Königsberg (today's Russian Kaliningrad).

Signed, Sealed, Delivered

In March 1736, Euler got a curious letter from his friend† Carl Leonhard Gottlieb Ehler, a mathematician, astronomer, and future mayor of Danzig in Prussia (today's picturesque coastal city of Gdańsk in Poland). Likely re-

*Only five of Euler's children survived beyond infancy.

†Mostly pen-friend, it seems.

ferring to an earlier conversation or a letter,* Ehler requested help—asking for a friend—with a seemingly simple problem:

> You would render to me and our friend Kühn a most valuable service, putting us greatly in your debt, most learned Sir, if you would send us the solution, which you know well, to the problem of the seven Königsberg bridges, together with a proof. It would prove to be an outstanding example of Calculi Situs [geometry of position], worthy of your great genius. I have added a sketch of the said bridges [. . .]

Heinrich Kühn, Ehler's protégé and professor of mathematics at the academic gymnasium in Danzig, must have been familiar with the folklore of his hometown, Königsberg. Like most European medieval settlements, this Prussian city was founded on the banks of a river. Königsberg spanned four land masses—two islands, Kneiphof and Lomse, and both banks of the Pregel—connected by seven bridges, as shown in Figure 5.1. According to the lore, locals enjoyed challenging themselves to stroll through the city's four sections crossing each bridge exactly once, not necessarily ending at the starting point, but nobody had found such a path yet. Did it even exist? Intrigued, Kühn introduced Euler to this puzzle, with Ehler acting as an intermediary.

Before you read on, try to find such a path through Königsberg yourself. It might sound easy. Indeed, Euler was initially unimpressed with the challenge, unclear why any mathematician would engage with such a trivial puzzle. He argued that "this type of solution bears little relationship to mathematics" and "the solution is based on reason alone, and its discovery does not depend on any mathematical principle."

Despite this skeptical initial response, just a few months later, Euler presented the solution to his colleagues at the Academy of Sciences in St. Petersburg, and in 1741 wrote the seminal paper "Solutio problematis ad geometriam situs pertinentis" ("The solution of a problem relating to the geometry of position"). In the paper, Euler referred to a branch of geometry that "is concerned only with the determination of position and its properties; it does not involve distances, nor calculations made with them."

*We know that the correspondence between Euler and Ehler lasted from 1735 to 1742, but the early letters didn't survive.

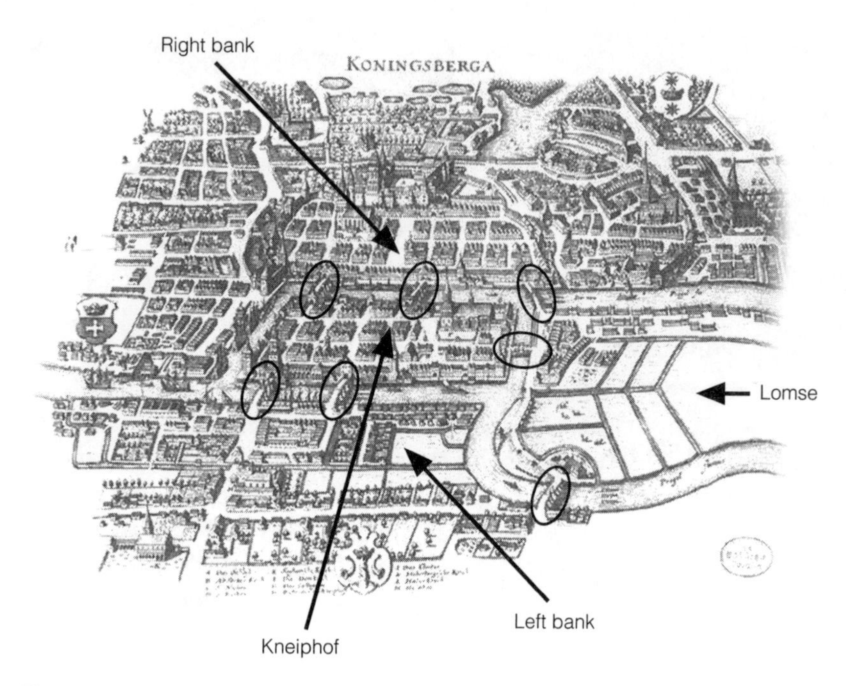

Figure 5.1 In Euler's time, the River Pregel divided Königsberg into four parts: two islands and two banks.

This description should remind you of Tube maps and topology, an area still in its infancy in Euler's times. In a way, Euler saw in Königsberg's map something that everyone else missed—or rather, he missed something that everyone saw: the unnecessary details.

This simple puzzle about Königsberg's seven bridges encouraged Euler to think about connections, which led to the invention of an ingenious mathematical tool: graphs. We've seen how a graph-based Tube map helps us navigate London; soon we'll discover some of graph theory's other applications. But first, let's solve the problem of Königsberg's seven bridges.

Bridging the Gap

Euler acknowledged that, in principle, one could list all possible walks in the city and check if they cross each bridge exactly once, but this would take too much work and time. So, just as Henry Beck transformed the metric plan of London into a topological Tube map, Euler stripped Königsberg's map of its distracting details.

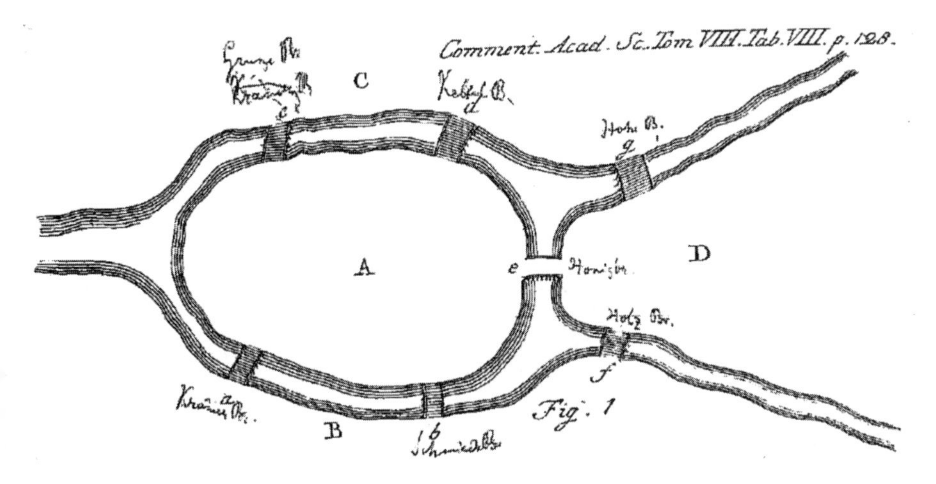

Figure 5.2 Euler removed unnecessary details from the map of Königsberg.

Euler figured that the shape of land masses didn't matter, so he redrew the map, labeling Kneiphof, the Pregel's left bank, the Pregel's right bank, and Lomse with letters A, B, C, and D, respectively. This way, he could describe the crossing of a bridge as a pair of letters—for example, AB meant getting from island A to bank B, regardless of the bridge crossed. Such pairs of letters were combined into longer sequences to describe whole walks—for example, ABDC meant going from A to B, from B to D, and from D to C.

To understand his reasoning, we'd need to carefully count how many times each letter appeared in a sequence describing a walk, which could get quite tiresome. Euler didn't have a choice, but we do—we can now represent his method visually with graphs, a tool that hadn't been invented yet in Euler's times. In modern terminology, Euler considered a graph—a set of vertices connected by edges—and wanted to find a path following each edge exactly once.

Our task is equivalent to a popular puzzle of drawing a figure—or a graph—without lifting the pen and without tracing any line more than once. Since we can use each edge only once, after we trace an edge, it's not available to us anymore. Let's consider what happens when we go through a vertex. First, we trace an edge to get to this vertex, and then we trace another edge to leave it, so passing through a vertex means using up two edges. The only exception to this rule is the vertex where we start tracing,

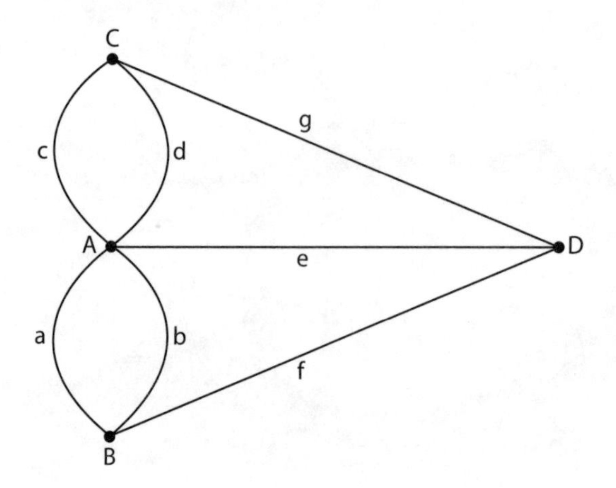

Figure 5.3 A graph representing landmasses and bridges in Königsberg.

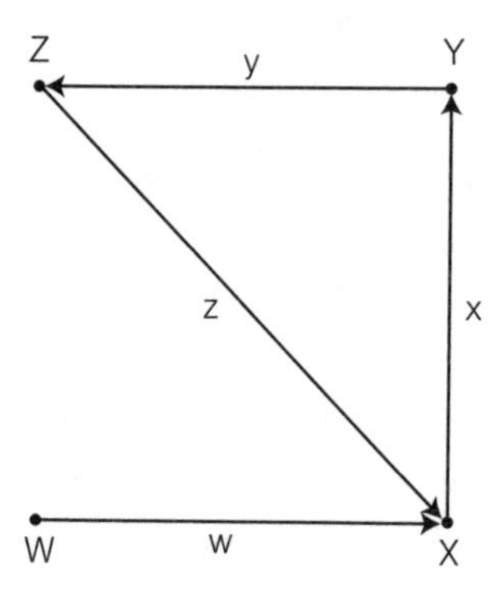

Figure 5.4 Whenever we pass through a vertex, we use up two edges: one to enter the vertex and one to leave it. The exceptions are the start point W, which we don't need to enter, and the end point X, which we don't need to leave.

as we don't need to use an edge to get there, and the vertex when we end tracing, as we don't need to use an edge to leave.

This is a crucial observation: it means that for any path to use all the edges exactly once, only two vertices can have an odd number of edges going through them—the start and end points. All other vertices must have an even number of edges, or, in graph lingo, an even degree. Euler noticed that in Königsberg, four land masses had an odd degree (although he didn't use this term), which meant that the required path didn't exist. No wonder nobody had found it!

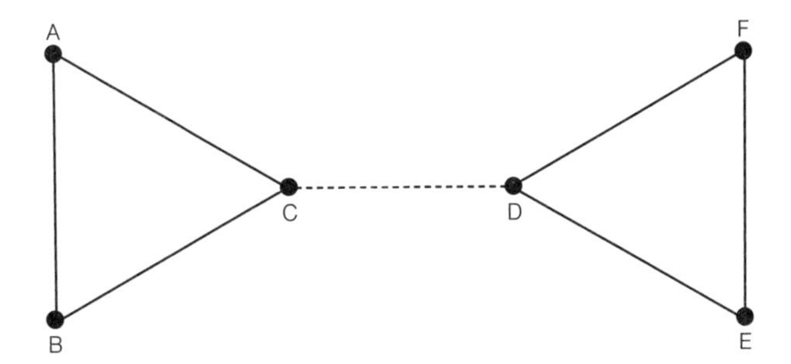

Figure 5.5 If we start from C, we cannot go to D, as removing the edge between C and D would separate the graph into two parts. An example of a path in this graph is CABCDEFD.

Euler also stated, without proof, the opposite—that if all vertices have an even degree, except possibly only two vertices, then it's possible to construct a path that passes each edge exactly once. Instead of going through the rather technical proof of this theorem, let's see how we can find such a path in practice. This simple, albeit inefficient method was first proposed in 1883 by French mathematician Pierre-Henry Fleury.

Given a graph, we first need to check the degrees of its vertices. If degrees are even, we can start at any vertex, if at most two are odd, we start at one of the odd vertices, and if more than two are odd, the path doesn't exist, so we give up. We travel from vertex to vertex, each time removing the "used" edge from the graph and writing down the path up to now. At each step, we can choose any neighboring vertex, with one caveat: if removing an edge would split the graph into two disconnected parts, we need to go to another vertex, as illustrated in Figure 5.5. This simple method will safely lead us through the last remaining edge, completing the path.

Euler not only answered the question the residents of Königsberg had posed but also found a simple way to establish whether it's possible to walk around any city crossing each bridge exactly once. Thanks to this general result, we can determine whether such a walk is possible in modern Kaliningrad, where only five bridges from the original problem remain, after two of the bridges were destroyed during the Second World War.*

* Only two have survived from Euler's time: two were replaced by a highway and one was rebuilt in 1935.

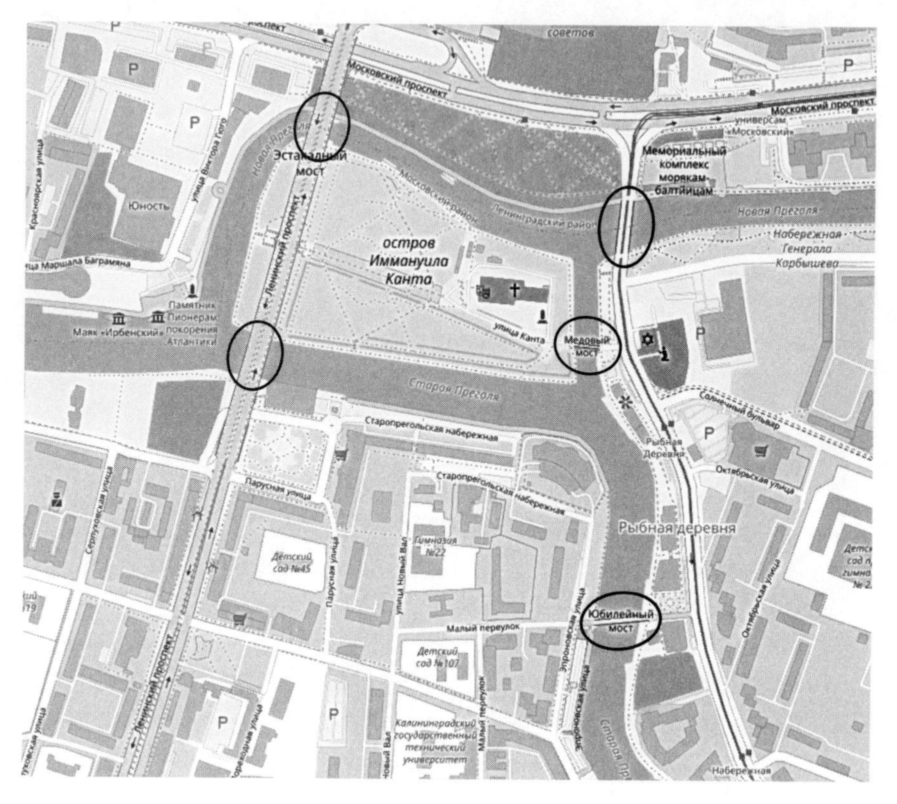

Figure 5.6 In today's Kaliningrad, only five of the seven bridges remain.

The loss of two bridges has a silver lining: today's residents of Kaliningrad can take a walk crossing each of the five bridges exactly once. We know it without even tracing their route: in a graph representation, both riverbanks have a degree two, and both islands a degree three. So, a route is possible as long as we start on one island and end on the other one.

The Birth of Graph Theory

At first glance, the Seven Bridges of Königsberg might seem to be a pastime for bored citizens of Königsberg, not worth the time of any serious mathematician. And it's not the puzzle that has gone down in history, but Euler's way of approaching the solution. His 1736 paper is considered to be the first in not one but two new fields of mathematics: topology, which we encountered in Chapter 4, and graph theory.

Graph theory is a branch of mathematics that deals with graphs—abstract structures consisting of finite sets of vertices, some connected with edges. Usually, we represent graphs as diagrams, depicting each vertex as a point and each edge as a line connecting two vertices. We've already seen such representations of graphs: for example, the representation of the topography of Königsberg, the London Tube map, and the Catawba Deerskin Map. While any book on graph theory is full of diagrams, the graphical representation is just a useful tool to study abstract mathematical structures. Although Euler neither drew a graph nor called his solution a graph,* he formulated the problem of the Seven Bridges of Königsberg in terms of vertices and edges. That's why we consider his paper the birth of graph theory, which was recognized as a separate branch of mathematics only in the twentieth century. The first textbook on this topic was published in 1936 by the Hungarian mathematician Dénes König, an aptly named researcher of the mathematical area inspired by Königsberg.

The Seven Bridges of Königsberg involved finding a tour of a city that passed every bridge in the city—or, in the language of graph theory, every edge in a graph—exactly once. A century later, an Irish mathematician, astronomer, and physicist, Sir William Rowan Hamilton, considered a seemingly similar, but in fact very different problem of visiting every *vertex* exactly once. Finding such a path turned out to be much, much harder than tracing every edge exactly once, which has very practical implications.

Packaged Mathematical Miracles

When the COVID-19 pandemic struck in early 2020, many stores closed and shopping in person wasn't safe, so delivery services became my best friend and likely yours too. During the lockdown, nothing would cheer me up more than a new book arriving at my door. These days, we take deliveries for granted, but each of them is a mathematical miracle.

Each day of 2021, the world's largest delivery company, United Parcel Service, delivered on average 25.2 million packages. According to Juan Perez, UPS's chief information and engineering officer, many UPS drivers stop as often as 135 times a day. Planning a route to so many destinations is unexpectedly challenging. In principle, one could list all possible routes,

*An English mathematician, James Joseph Sylvester, introduced the term *graph* only in 1878, comparing it to graphical representations of chemical molecules.

compute their lengths, and pick the shortest* one, but this would take a long, long time.

If a driver had only three deliveries to make, she could first deliver any item, then any of the remaining two, which would leave the final package to deliver. In total, she could plan $3 \times 2 \times 1 = 6$ different routes.† Through similar reasoning, we can figure that with five deliveries, there are $5 \times 4 \times 3 \times 2 \times 1 = 120$ possible routes; with ten, there are $10 \times 9 \times 8 \times 7 \times 6 \times 5 \times 4 \times 3 \times 2 \times 1 = 3{,}628{,}800$ possible routes, and with 135 deliveries, there are way more possibilities than stars in the universe, cells in the bodies of all humans, and trees on the Earth put together—and these are the possible routes of *one* of the thousands of UPS drivers on *one* day! Without a smarter route planning system, UPS's first delivery person would be still planning his first delivery, over a century after UPS was established.

A Troubled Salesperson

Optimal route scheduling problems predate UPS and other modern delivery services. For centuries, hawkers would go from door to door, village to village, selling anything from apples to coal to hats. In the nineteenth century, companies mushrooming in Europe and the United States turned this generally disrespected activity into an important profession. To travel efficiently all over the country advertising and selling products, salespeople would consult handbooks such as the 1832 German *Der Handlungsreisende—wie er sein soll und was er zu tun hat, um Aufträge zu erhalten und eines glücklichen Erfolgs in seinen Geschäften gewiß zu sein—von einem alten Commis-Voyageur* (*The Traveling Salesman—How He Should Be and What He Has to Do, to Obtain Orders and to Be Sure of a Happy Success in His Business—by an Old Traveling Salesman*), which is considered the first recorded mention of the route planning problem. This snappily titled booklet specified that "the main point [of the optimal route] always consists of visiting as many places as possible, without having to touch the same place twice." The author suggested five possible tours through forty-five German

* As we saw in the previous chapter, the shortest route might be different depending on the type of distance we care about. We should define the distance we want to minimize, for example, the Euclidean distance or the average driving time.

† If you're not convinced, imagine that the three destinations are A, B, and C. Then, the possible routes are ABC, ACB, BAC, BCA, CAB, and CBA, which makes six routes in total.

cities based on trial and error rather than rigorous mathematical reasoning. It took the passing of almost another century for the first mathematical formulation of the salesperson's problem to appear, and it did so in the work of Viennese mathematician Karl Menger.*

One of Menger's main research interests was measuring lengths of curves in space, which—as we saw in Chapter 3—is trickier than it sounds. In the 1920s, he enjoyed discussing optimal routes with his colleagues at the University of Vienna. At the Vienna Mathematics Colloquium in 1930, he presented the "messenger problem" faced by postal workers and travelers alike: "given a finite number of points with known pairwise distances, to find the shortest path connecting the points." In other words, Menger wanted to find a path visiting every point exactly once, but he made this already hard problem even more complex by requesting not *any* path, but the *shortest* one. He was aware of its difficulty and suggested a way of finding approximate solutions, noting that it might not always lead to *the* shortest path.

By the 1940s, the routing problem was mentioned in different contexts under different names. At some point, someone called it the "traveling salesman problem" (TSP), and the name stuck.† Its first recorded use comes from a 1949 research paper by an exceptional mathematician, Julia Robinson,‡ although the mathematical community must have already been familiar with the term.

Robinson spent about a year at the RAND Corporation, an American research and development institution founded soon after the Second World War to support the US Armed Forces.§ When faced with particularly

*Karl Menger is probably best known for Menger's sponge, a fractal resembling a Rubik's Cube with infinitely many holes.

†In the rest of the chapter, I'll be using my preferred gender-neutral alternative, "traveling salesperson problem."

‡Julia Robinson's career illustrates the obstacles faced by women in mathematics. After marrying a fellow mathematician, Raphael M. Robinson, she was unable to teach in the mathematics department at the University of California, Berkeley, because of the nepotism rule. She had to teach statistics instead, which didn't align with her interests. Even after her husband retired, it took the department a few years to offer her a full-time professorship, and it only happened after she became the first female member of the National Academy of Sciences. Julia also recalled smaller but not less painful acts of discrimination: "Both my husband and I were invited to a conference and the committee decided it would be unfair to pay expenses for both of us because the other families would have to pay for the wives." All these obstacles make Julia Robinson's mathematical achievements even more exceptional.

§Today RAND's researchers study a variety of topics, from health to sustainability, to energy to education. They even run a public policy PhD program.

challenging problems, the organization would offer prizes for solutions. Nobody ever received the prize for solving the TSP, but the bounty surely encouraged researchers to have a go at tackling the problem. Since then, we've managed to find shortest routes between larger and larger sets of destinations, but the solution to the general TSP continues to elude us.

Which begs the question, what if the shortest route isn't what we should be searching for?

Good Enough

Over decades, salespeople have been touring the world unaware of mathematicians grappling with the traveling salesperson problem. Without the optimal solution, travelers make do with good-enough itineraries. Inspired by this practical approach, researchers have developed a whole range of heuristics—from Greek *heuriskein* ("to find, to discover")—which are methods of finding an approximate solution when solving the problem exactly would be too hard, too slow, or simply impossible. For TSP, it means finding a tour that might not be the shortest possible but short enough. Researchers have developed many such rules of thumb that provide solutions to the TSP without guaranteeing their optimality.

A heuristic is a type of algorithm, which is a sequence of well-defined instructions.* The word algorithm is mostly used by mathematicians and computer scientists, but baking recipes are also algorithms: they specify ingredients and their precise quantities, and describe exactly what one needs to do with them in what order to get a delicious banana bread or apple pie. Some algorithms are exact, which means that they always lead to the optimal solution but, in the case of the TSP, an exact algorithm would be impractical. If you recall the UPS problem, you'll understand why we settle for imperfect results; even for the relatively small numbers of places the traveling salesperson must visit, the number of possible paths is so large that exploring them all is not an option. So, we need to use heuristics, such as the simple "nearest neighbor" algorithm.

*The relationship between heuristics and algorithms is more subtle. Some heuristics are types of algorithms, while some are contrasted with algorithms. The heuristics I describe in this chapter have well-specified instructions, although they might not lead to the optimal solution. That's why I categorize them as algorithms.

Nearest Neighbors

The nearest neighbor algorithm is as intuitive as it gets: from the starting point, go to the closest destination and then always move to the closest destination that you haven't yet visited. We call this type of algorithm "greedy" because it prioritizes the best option at the moment, without considering the long-term outcome.

Sometimes this approach gives a satisfactory result—maybe not the shortest route, but not too far from it either, like in Figure 5.7. Sometimes, however, it fails quite badly. The algorithm might be doing well until the last destination, but then the salesperson will need to return to the starting point, no matter how far away it now is. Always following what feels good right now is a good approach to neither life nor route optimization. If the distances between locations follow the triangle inequality,* the situation is a bit better as it limits how bad the nearest neighbor route might get. In this case, this heuristic won't give the average UPS driver who needs to visit 135 places in a working day a route longer than about 4.5 times the shortest one.† Sure, it's not ideal, but it's still surprisingly good given the speed of computations.

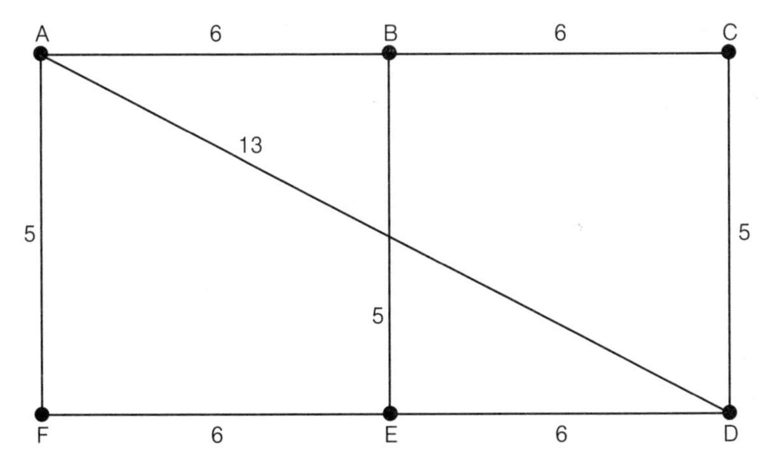

Figure 5.7 The nearest neighbor route AFEBCDA has length 5+6+5+6+5+13=40, which is longer than the optimal route ABCDEFA of length 6+6+5+6+6+5=34.

*Remember, this means that the direct journey, without any deviations, gives the shortest distance between two destinations.

†More generally, the nearest neighbor solution to a TSP problem with n places (again, assuming symmetric distances and triangle inequality) won't be worse than $1+0.5(\log_2 n)$ times the true optimal solution.

Meals on Wheels

The nearest neighbor algorithm works well, but in some situations, computations still might take too long, especially when the locations to visit change frequently. To address this problem, four decades ago researchers developed another clever heuristic.

Volunteers and employees of Meals on Wheels regularly deliver lunches to people unable to buy or prepare their meals for themselves, primarily senior citizens. In the 1980s, the manager of this nonprofit organization running the program in Atlanta, Georgia, was responsible for everything from planning menus and ordering meals to maintaining the list of clients. On top of that, every day she would hand each of the four drivers a list of thirty to forty locations they had to deliver hot lunches to within four hours, together with a suggested route. This is a perfect example of the traveling salesperson problem, which in principle could be solved with a heuristic such as the nearest neighbor. But there's a caveat: the list of clients changed frequently. Someone would get better and not need the service anymore, someone else would get sick and suddenly require help or, sadly, someone would die. This required the already overworked manager to replan the route day in, day out. The underfunded Meals on Wheels service couldn't afford to hire another employee, not to mention access any reasonable computational power to help with the constant rerouting. The manager's ordeal lasted until scientists at the Georgia Institute of Technology came up with an ingenious solution.

John J. Bartholdi III and colleagues designed a heuristic that allowed fast, cheap, and low-tech routing, easy to adjust when the list of clients changed. This simple system was based on the complex mathematical idea we encountered in Chapter 3: the space-filling curve. Recall that a space-filling curve is an infinitely long line that fills a whole two-dimensional space, which we can also think about as a possible route for a person who needs to visit every single point in a given region. Of course, the Meals on Wheels driver had to visit only a subset of locations and not *every* point in Atlanta, but because a space-filling curve encompasses all points in the region, each delivery location has a unique position along the curve. In other words, the curve prescribes the order in which the clients should be visited.

Instead of a supercomputer, the system designed at Georgia Tech relied on an Atlanta street map with a clear coordinate grid, a table relating the

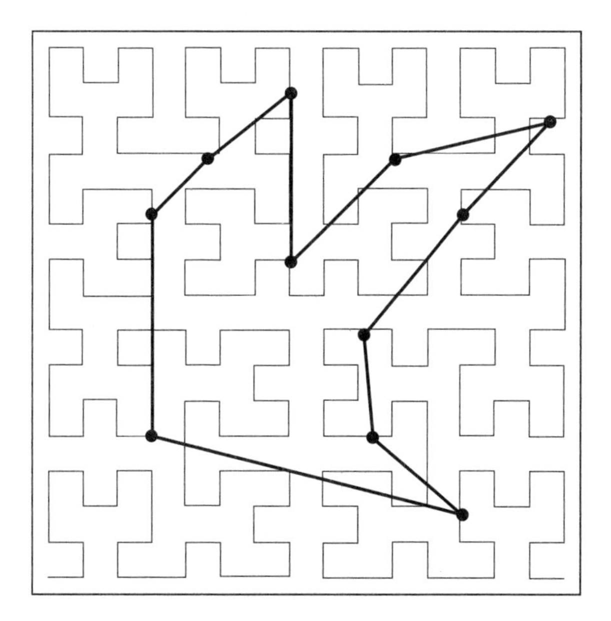

Figure 5.8 The meals should be delivered in the order represented by the infinite version of this curve. The dots represent the locations of clients.

coordinates to their position along the curve, and two Rolodex card files.* To create the table, the researchers converted each pair of coordinates (horizontal and vertical) from the map into a single number representing its relative distance from the beginning of the curve, a value which uniquely identified a delivery location. Both Rolodex files contained cards corresponding to current clients: one sorted alphabetically by clients' names and one sorted numerically by their location along the space-filling curve. The latter file was divided into four roughly equal parts, each assigned to one of the four drivers. All each driver needed to do was to deliver the meals in the order provided by their part of the Rolodex.

This system allowed the manager to easily adjust the delivery routes without any need for time-consuming recalculation. Whenever a new client appeared, the manager would simply read their location from the map, convert the coordinates to the position along the curve using the table, and insert two new cards into the two Rolodexes. If someone didn't need lunch

* Rolodex is a business card storage device, which allows the user to quickly navigate to a specific card by rolling them (hence the name: rolling + index).

deliveries anymore, both of their cards would be removed. On average, this simple, intuitive routing system created routes longer than the shortest route by about 25 percent, but the gains in the manager's time (and sanity) more than made up for this imperfection.

Ant Colonies

If you've ever observed an anthill, you've surely noticed swarms of ants leaving the nest with purpose and returning with pieces of food often bigger than their little bodies. For these insects, time and energy are even more precious than for UPS drivers, so they've evolved to seek shortest paths between the food source and the nest. But how do these little creatures achieve that without computers and human brains?

When an ant leaves the nest, it doesn't know where the food is, so it picks a direction at random. As it walks, it leaves behind a trail of pheromones, the chemicals released by animals that alter the behavior of other individuals of the same species. When an ant finds food and walks back to the nest, it leaves twice as much pheromone as another ant that has walked the same distance in another direction but didn't find any food. When the next ant leaves the nest, it doesn't have to pick the route completely at random but chooses the one with the largest amount of pheromone. The shorter the route to the food source, the faster the ant will get there and back, leaving more pheromones along the way, thereby strengthening this route. This way, the ants will eventually stop taking inefficient paths and will all follow the same optimal route (or routes), which some of us learned the hard way by discovering an army of ants efficiently marching toward cookie crumbs accidentally left in the kitchen. This natural method of following shortest paths is flexible—if an obstacle appears along a route, about half of the ants turn right and half of them turn left to avoid it, but then pheromone trails quickly teach them which new path gets them to food more quickly.

In the 1950s, American inventor and bioengineer Otto Schmitt described the idea of biomimetics, a term derived from the Ancient Greek *bios* ("life") and *mīmēsis* ("imitation"). He might have been the first to name the idea, but he was not the first to apply it. For example, Leonardo da Vinci understood that instead of inventing a flying machine from scratch, he could study the anatomy of birds and copy nature's solutions. Nature has had millions of years to develop optimal solutions to a variety of problems, and it doesn't care about copyright infringement, so why not imitate it? Following this idea, in his 1992 doctoral thesis, an Italian engineer, Marco Dorigo, proposed the first ant colony optimization algorithm based on ants communicating with pheromones. He imagined replacing the salesperson with a colony of ants and the salesperson's destinations, represented by vertices of a graph, with food sources. The vertices were connected by edges with numbers corresponding to the amount of "pheromone" left by the "ants" traveling through the graph. In the end, both ants and travelers have a similar goal: to get to the destination as fast as possible.

By mimicking nature's ingenuity, we can efficiently find approximate solutions to the traveling salesperson problem. Since the publication of Dorigo's thesis, improved variations of his ant colony optimization algorithm have been applied to problems ranging from bankruptcy prediction to power electronic circuit design.

No Left Turns

We've looked at three examples of simple heuristics that provide a reasonable balance between the speed of computation and the length of the route found. Even if the result isn't truly optimal, for many applications it suffices. However, for a company as huge as UPS, even tiny differences in delivery time quickly add up and start to cost serious money, so they've invested a lot of resources to improve their routes. Efficient route planning is a considerable component of the company's success, so technical details are top secret, but this didn't stop me from trying to piece together the publicly available information and work out how they do it.

Surprisingly, the biggest challenge of delivery services isn't sending the package to the other side of the world, but the so-called "last mile delivery," which is getting the goods from the final hub to the recipient. For decades, UPS was small enough to deal with it in an ad hoc manner, but after its rapid growth in the 1960s and 1970s, it required some careful planning. So, they kept developing the industrial engineering department, to which the responsibility of route planning fell.

Back then, UPS faced similar problems to Atlanta's Meals on Wheels team—the lack of internet and mobile phones, as well as constantly changing destinations—albeit on a much bigger scale. The area around each hub would be divided into smaller segments, each in the realm of responsibility of one driver. In each segment an industrial engineer would draw a base route that covered all possible locations, mimicking the idea of space-filling curves but accounting for other constraints, such as serving businesses during working hours or avoiding the busiest traffic. On the day, the driver would adjust her or his base route depending on the packages to be delivered.

At the dawn of the new millennium, the growing popularity of delivery services and the introduction of delivery time windows and other premium options made this pen-and-paper, driver-reliant planning system insufficient. In 2003, UPS started a long and ambitious project called On-Road Integrated Optimization and Navigation (ORION). At first, the team of UPS mathematicians and engineers approached the task as a traveling salesperson problem, additionally accounting for delivery time windows. They estimated that if each UPS driver in the United States was idle for one minute each day, after a year the losses would add up to fifteen million dollars, so their algorithm had to find routes almost instantly. They spent months developing a new heuristic, which unfortunately didn't pass extensive tests. Given a twelve-month deadline from management to come up with something better, the team started analyzing what had gone wrong.

ORION's goal was to automate the routing process, but it turned out that taking the drivers out of the equation hadn't been the best move. Over the years, drivers develop delivery patterns based on factors ignored by the algorithm. For example, they avoid the neighborhood of a local school in the afternoon when hundreds of students spill out into the streets. Also, the initial version of ORION didn't consider the drivers' safety, so the improved product preferred routes that avoided dangerous left turns* and changing lanes.

The researchers could have explicitly implemented all possible constraints (such as avoiding the end of the school day), but they decided to trust the drivers' experience. They designed the algorithm so that it preferred routes close to routes taken by UPS drivers in the past. This way, they could rely on human experience and intuition without burdening the drivers with daily decisions. While the details of the final implementation are a secret, we know that the successful algorithm consists of two main parts: the

*In the United Kingdom and other countries with left-hand side traffic, this would correspond to right turns.

information about the routes historically preferred by the drivers and the optimal solution to a modified traveling salesperson problem.

According to UPS's estimates, since its deployment in 2012, ORION has been saving them more money annually than its development cost in total—and we're talking hundreds of millions of dollars per year. But what's more important to us all is that more efficient routes mean less pollution and carbon emissions. Now, the improved algorithm uses machine learning and GPS data to continuously update optimal routes based on changing traffic and delivery locations.

Without algorithms and powerful computers, we'd be helpless when faced with a task as simple as choosing an optimal route. A human lifetime is too short to comb through the astronomical number of possible paths, even through relatively few points. By turning maps into graphs, we can convert questions about maps and routes into a language that can be understood by a computer. This trick, which we can trace back to Euler, allows us to tackle questions that otherwise would be too complex to answer. Next, we'll see that answering one of them took centuries, but the attempt brought us new methods of tackling real-life problems.

Wedding Planning

When Marlena, one of my closest friends, announced her engagement, we both knew that the months leading up to her wedding would be stressful. Invitations, catering, photographers . . . it's a lot to deal with. But the biggest challenge arose after the venue was chosen, the first dance choreographed, and the cake tasted: designing the seating plan turned out to be a nightmare. How to make sure that the aunt doesn't share a table with the uncle whom she'd divorced a decade ago? How to separate the grandma who doesn't speak a word of English from British university friends?

Although I have no idea about wedding planning, this was something I could help with by using graphs. To simplify the problem, let's say that Marlena invited five guests: Vicky, Will, Xena, Yara, and Zack. Vicky can't stand Will, Yara, and Zack; Will is Xena's ex; Xena and Yara always find a reason to argue; and Yara is Zack's biggest enemy. We can illustrate this problem with a graph, where vertices correspond to the five guests and the ones who shouldn't sit together are connected by edges, as in Figure 5.9.

We'll mark the vertices so that if two guests are assigned the same letter, they'll share the table, and they won't share the table otherwise. Then, our

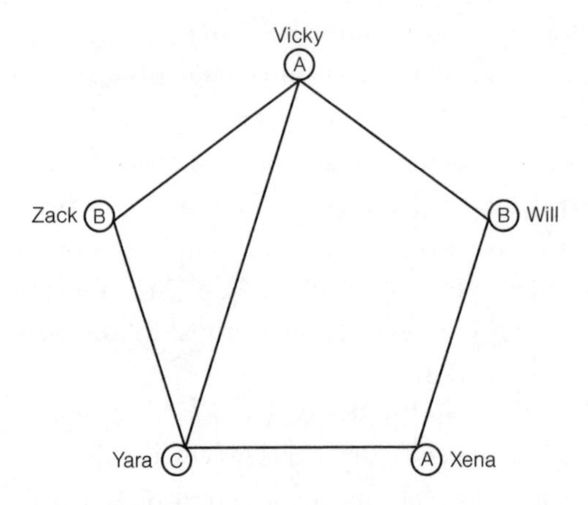

Figure 5.9 In this graph, vertices representing party guests who shouldn't sit together are connected by edges. Each letter represents a different table. Guests labeled with the same letter should sit together.

task is to label all vertices so that no two neighbors—vertices sharing an edge—are labeled with the same letter. Of course, we could assign each vertex a different letter, but a wedding with each guest seated at a separate table wouldn't be much fun. Besides, there's only a limited number of tables that the wedding venue can hold. So, what's the *smallest* number of tables needed to seat these five guests? In other words, how many letters do we need?

Let's start by labeling Vicky "A." Will shares an edge with Vicky, so he needs to get a different label—let's label him "B." Zack also shares an edge with Vicky but doesn't share an edge with Will, so he also can get B. Now, Xena cannot be B but can be A since she doesn't share an edge with Vicky. Finally, Yara shares an edge both with Vicky and Zack, so she needs a third letter: C. Figure 5.9 shows that these five guests need to sit at three separate tables: Vicky and Xena at A, Will and Zack at B, and Yara at C.

We've solved this example by hand, and although we'd need a computer to design a seating plan for a real wedding, the method would stay the same. Graph labeling (or coloring, if you're not restricted by black-and-white printing!) is useful whenever we need to avoid conflicts and clashes—between wedding guests, radio frequencies, or even aircraft crews who cannot be on two planes at the same time. Colorful graphs make these real-world problems not only more visually appealing and intuitive but also easy to communicate to a computer. It's hard to believe that all these wide-ranging applications have a single source, and a rather childish one: map coloring.

Four Colors Suffice?

In the mid-1850s, Frederick Guthrie, one of the future founders of the Physical Society in London but a student at the time, wondered if and why any figure divided into regions can be colored with at most four colors, so that no two regions sharing a boundary have the same color. His question was inspired by his older brother, Francis, who had made this observation while coloring the map of English counties. I wonder why a South African who would become a successful mathematician and botanist, and an eponym for several plant species, deemed map coloring a worthwhile activity, but we can only be glad that he did.

The four-color conjecture, as the problem was later named, involves any map divided into regions—for example, counties, states, or countries—some of which share boundaries. The conjecture states that to color *any* such map so that no two neighboring regions share the same color we need *at most* four colors. In the rest of the chapter, by a map I mean any figure divided into regions which I'll call countries, but the conjecture isn't limited to the political maps we find in atlases. If the four-color conjecture is true, it needs to hold for the map of L. F. Baum's Land of Oz, J. R. R. Tolkien's Middle-earth, or any zigzags you can draw on a sheet of paper. Again, in the absence of color printing, I've substituted different letters for different colors in the figures that follow.

Two countries are considered neighbors if they share a boundary, but not if they meet only at a single point—for example, Utah and New Mexico can have the same color (or letter in our illustration), and so can Arizona and Colorado because they are connected only in one corner, rather than along the border, as in Figure 5.10.* Another important restriction is that all regions must be connected, so, for our purposes, exclaves aren't considered parts of the main country—for example, Gibraltar and mainland Great Britain could get different colors. Frederick Guthrie had noticed that the

*If you want to visit these four US states at the same time, you need to head to the Four Corners Monument.

Figure 5.10 Utah and New Mexico can have the same letter, and so can Colorado and Arizona, but these two letters cannot be the same.

four-color conjecture refers only to two-dimensional surfaces, such as flat maps and the surface of a sphere; in three dimensions, one could easily construct a map that requires any number of colors, like the five-country "map" in Figure 5.11. So, we'll only consider two-dimensional maps.

The four-color conjecture intrigued the Guthrie brothers' teacher, Augustus De Morgan, who happened to be one of the most famous British mathematicians and logicians.* He created an example of a map requiring four colors—similar to the one on the left in Figure 5.12—immediately disproving a potential three-color conjecture. A real-world example could be a political map of South America, as shown on the right in Figure 5.12, because of Paraguay—a landlocked country surrounded by Argentina, Bolivia, and Brazil, each bordering the other two. Each of these four countries needs a different color, so we cannot color South America with fewer than four.

But can a map require more than four colors? This question baffled De Morgan. In October 1852 he wrote a letter to Sir William Rowan Hamilton, the Irish mathematician who studied the UPS problem before UPS

*Students of logic will recognize De Morgan's name from De Morgan's laws, which describe the rules of negating logical statements.

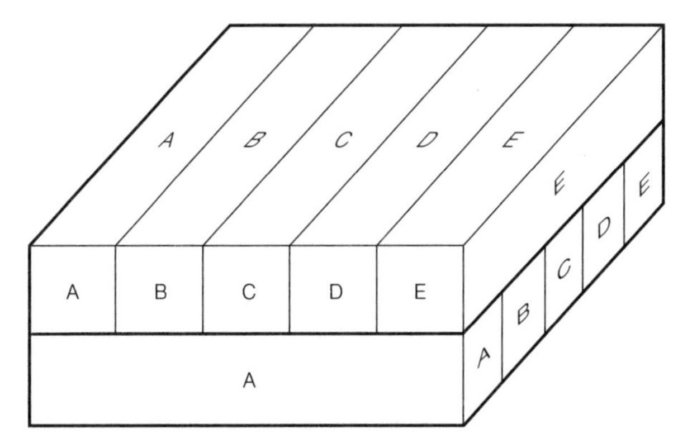

Figure 5.11 An example of a three-dimensional "map" requiring five colors constructed by Austrian mathematician Heinrich Tietze.

Figure 5.12 Maps requiring four colors: De Morgan's example (*left*) and the political map of South America (*right*), where Paraguay is the landlocked country labeled B.

was founded. The problem seemed so simple that De Morgan warned the addressee: "If you retort with some very simple case which makes me out a stupid animal, I think I must do as the Sphynx did . . ."* Hamilton didn't prove or disprove the four-color conjecture; he didn't show any interest in

*In an ancient Greek myth, the Sphinx killed herself after Oedipus solved the riddle that she had posed.

it whatsoever. "I am not likely to attempt your 'quaternion'* of colours very soon," he responded dismissively, thus missing out on an opportunity to participate in what became an exciting quest.

Why Is It So Hard?

Augustus De Morgan didn't live to see the proof of the four-color conjecture. A few years after he had died in 1871, the problem was picked up by another star of mathematics, Arthur Cayley. Having obtained a mathematics degree at Trinity College, Cambridge, with the top overall mark, he became the youngest person that century to be offered a Cambridge Fellowship. However, reluctant to fulfill the religious requirements of the position, Cayley refused the honor and became a successful lawyer instead, although he continued his research in mathematics throughout his legal practice. At last, in 1863 he was offered an Honorary Fellowship at Trinity College, this time with religious exemption—and this is where he spent the rest of his life.

At an 1878 meeting of the prestigious London Mathematical Society, Cayley inquired whether "the statement that in colouring a map of a country, divided into counties, only four colours are required, so that no two adjacent counties should be painted in the same colour" had been solved. It hadn't. So, Cayley picked up the baton from the late De Morgan and the following year published a paper explaining why proving the map-coloring conjecture was so difficult. Interestingly, the article appeared not in a mathematical journal, but in the newly established *Proceedings of the Royal Geographical Society,* thus connecting math and maps before it was cool. Cayley observed that if we add another country to a four-colored map, it's not clear whether we'll be able to keep the initial coloring or whether we'll have to start from scratch. In particular, he noticed that if all four colors appear at the boundary, the addition of a new country surrounding the previous area would require recoloring the map. In other words, we might need to recolor the whole map every time we add a new country, which makes finding some general coloring pattern difficult, if not impossible.

Although Cayley didn't manage to prove or disprove the four-color conjecture, his interest in the problem and valuable insights moved the research forward. Not long after Cayley's paper for the Royal Geographical Society,

*One of Hamilton's many achievements was creating a new number system based on special expressions called quaternions.

Figure 5.13 We can convert colored maps to colored graphs.

readers of the prestigious journal *Nature* were in for a real treat: a British barrister and Cayley's ex-student, Alfred Kempe, announced that he'd proved the four-color conjecture. We call a proven conjecture a theorem, so Kempe turned the four-color conjecture into the four-color theorem—but not for long.

Kempe suggested drawing one point in each country and connecting points corresponding to countries sharing boundaries to form a "linkage," as shown in Figure 5.13. We've already seen this structure—Kempe's "linkages" were graphs. More precisely, they were planar graphs, which are graphs that can be drawn on a plane (a flat surface) with no edges crossing. So instead of having to deal with a map, we can consider a graph whose vertices correspond to the countries and whose edges connect neighboring countries.

This observation was a game changer. To color a map, we don't have to worry about the country's shape and size, only about who its neighbors are. Graphs let us strip the problem of distracting details and focus on the most important stuff. This way, the coloring problem could be stated as a question of whether one could color all vertices of a graph drawn on a plane with at most four colors so that no two vertices connected by a single edge had the same color. This idea turned out to have applications beyond proving the four-color conjecture—such as solving Marlena's wedding seating conundrum.

Make It Five

The four-color theorem enjoyed its proven status for over a decade until another British mathematician named Percy Heawood found an irreparable flaw in Kempe's argument.* Although the four-color theorem reverted to

*Kempe's proof relied on complicated patterns of recoloring the countries, and Heawood discovered that in a particular case of a country with five neighbors, Kempe's method might not work. The mistake was so subtle that for most maps, it wouldn't matter. To show that Kempe's argument wasn't foolproof, Heawood had to construct a complex map containing dozens of countries.

being the four-color conjecture, Kempe's insights led to significant progress in proving the stubborn problem. As mathematician and historian of mathematics Robin Wilson phrased it, "it was incorrect, but it was a very good incorrect proof."

Heawood used Kempe's method to prove another powerful, albeit slightly weaker statement: the five-color theorem. Unsurprisingly, this theorem states that all maps can be colored with only *five* colors so that neighboring countries have different colors. Heawood's proof, written in nineteenth-century language, isn't a light read, so I'll modify it using modern concepts. We'll consider a graph drawn on a plane representing the map rather than the map itself, to focus on the most important details. As you read the next few paragraphs, remember that proving the five-color theorem took professional mathematicians decades. It's not supposed to feel easy, but getting a general idea about the method will help us appreciate the eventual successful proof of the four-color theorem and the diverse applications of graph coloring.

We'll use a rather unintuitive method of proving theorems called *proof by contradiction.* In this technique, we initially assume that the statement we're proving to be true is false. Then, we show that this leads to absurd consequences and conclude that our initial assumption must have been incorrect. This way, we indirectly prove that our hypothesis is true.*

To prove that every graph can be colored with at most five colors, we first assume that the opposite is true, that there exists a graph that needs *at least* six colors. We hope that this assumption will lead to some contradiction, which would prove our assumption wrong and, in turn, prove the theorem. From all possible graphs that need at least six colors, we pick the one with the smallest number of vertices and call it the *minimal counterexample.* In other words, the minimal counterexample is a graph that *cannot* be colored with five colors, but any graph with fewer vertices can.

We pick and remove any vertex—let's call it V—with at most five neighbors.† The new graph has fewer vertices than our minimal counterexample, so by definition of the *minimal* counterexample, it can be colored with five

* Proof by contradiction has its roots in the rule of logic that "if A is true, then B is true" is equivalent to "if B is false, then A is false."

† We can prove that every planar graph contains such a vertex using a tool called the Euler characteristic. I told you that Euler had achieved a lot!

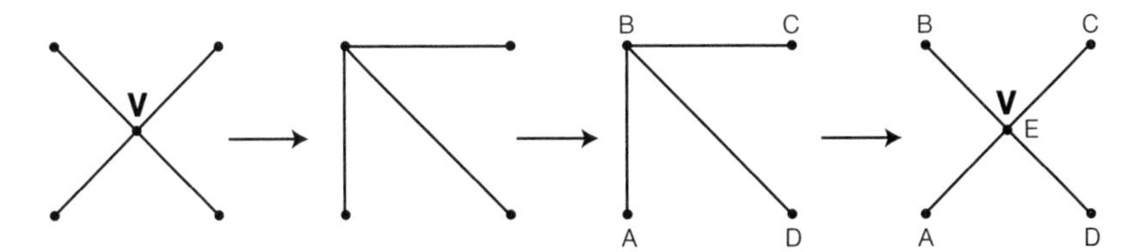

Figure 5.14 An example of a vertex V with at most five neighbors. When we remove V from the graph, we can color the remaining vertices using four colors. Then, the fifth color remains for vertex V.

colors. Now we need to bring V back in and show that the bigger graph also can be colored with five colors, as in Figure 5.14.

If the neighbors of V are colored with fewer than five colors, then we simply color V with the remaining color, and we're done. If, however, the neighbors of V already use all five colors, we need to do some reshuffling. I'm going to spare you the technical details, but Heawood showed that for any graph drawn on a plane, it's possible to recolor these neighbors so that they use only four colors. Then, we color V with the remaining one. In any case, we were able to color the minimal counterexample with only five colors, which means that it wasn't a counterexample after all. This way, we proved the five-color theorem by contradiction.

Why can't we similarly prove the four-color conjecture? For this argument to work, we had to remove a vertex with at most five neighbors. To prove the four-color theorem this way, however, we'd need a vertex with at most *four* neighbors. While we can prove that every planar graph contains a vertex with at most five neighbors, some planar graphs contain *only* vertices with more than four neighbors. So, this line of reasoning wouldn't work for the four-color conjecture.

The Building Blocks

At the turn of the twentieth century, not much progress had been made since Kempe's and Heawood's work. Finally, the topic was picked up by mathematicians in the United States, who expanded on ideas from Kempe's faulty proof. In general, they preferred to work on graph representations rather than actual maps—as we've seen, these representations are equivalent, but graphs are more convenient.

While proving the five-color theorem, we saw that *every* graph drawn on a plane must contain a vertex with one, two, three, four, or five neighbors.

When we remove this vertex (in our example it was called V), color the rest of the graph, and put it back in, we'll be able to complete the coloring. Over time, the researchers' goal became to find vertices with a similar property: if the graph can be four-colored when this vertex isn't there, it can also be four-colored with this vertex included. Additionally, this type of vertex must appear in every single graph.* This requirement made the task extremely difficult, since there are many possible graphs (or, equivalently, maps), and mathematicians had to make sure that such a vertex appears in *all* of them. Remember that we're not considering only existing real-world maps, but all possible combinations of neighboring regions.

Four Colors Suffice!

In the late 1940s, a twenty-year-old student of mathematics, philosophy, and physics at the University of Kiel in Germany, Wolfgang Haken, learned about three unsolved problems that had puzzled mathematicians for decades: the four-color conjecture, the knot problem of determining if a given string in three dimensions can be unknotted, and the Poincaré conjecture, a topological problem solved only in 2002 by an eccentric Russian mathematician named Grigori Perelman. Ambitious Haken was determined to tackle all three longstanding problems. He managed to solve the knot problem and spent over a decade dealing with hundreds of cases of the Poincaré conjecture. Finally, he turned his attention to the four-color conjecture.

Haken quickly understood that the problem couldn't be solved without a computer. Frustrated by his limited programming skills and the hesitancy of the programmers he had asked for help, he announced at one of his lectures that he was quitting his endeavors. It was a lucky coincidence that Kenneth Appel, a mathematician and a top-notch programmer, was sitting in the audience. He later approached Haken, saying: "I don't know of anything involving computers that can't be done; some things just take longer than others. Why don't we take a shot at it?" In late 1972, Appel and Haken began their successful collaboration. Two years later, they realized they

* It's a bit more complicated than that. The goal was to find a set of configurations of vertices such that any graph contains at least one of these configurations (a so-called "unavoidable set"). Each of these configurations must have the following property: if a graph can be four-colored with this configuration removed, it can also be four-colored with this configuration included (we call such configurations "reducible").

wouldn't be able to finish alone, and were joined by a graduate computer science student, John Koch. The last stretch of the proof involved a lot of computing effort, and the University of Illinois computer center was exceptionally supportive of their project, letting them use powerful computers during downtime. Appel and Haken later admitted: "We now know that this policy was essential to our success."

The computer sped up the process, but the researchers still had to manually check thousands of possible configurations of vertices. Haken asked his eldest, then-teenage daughter Dorothea—today a computer scientist, Dorothea Blostein—to help with this mundane job. And then, in June 1976, they found the holy grail: a configuration of vertices with the desired properties that appears in every possible map.* Appel wrote on the department's blackboard: "Modulo† careful checking, it appears that four colors suffice."

The "careful checking" had to happen quickly, as Appel was shortly off on a sabbatical, and a few other teams seemed to be getting close to a solution—which is how Appel's and Haken's children got their unusual summer jobs as proof checkers. When I speak to Blostein about this experience, she recalls sitting in the majestic Altgeld Hall of the Department of Mathematics at the University of Illinois Urbana-Champaign, where sounds from the bell tower interrupted their work every hour. Her job was to go through stacks of computer printouts generated by Appel's programs and check if the printed and hand-drawn configurations matched.

Appel's and Haken's children, checking hundreds of pages of the proof by hand, found quite a few mistakes, but all were easily repaired. Appel and Haken understood even more: even if some mistakes had been missed, they'd be able to repair them, which made the proof exceptionally robust. So, on July 22, 1976, they announced that the four-color conjecture had become the four-color theorem.

What Is a Proof?

The scientific community's reactions to this breakthrough varied. On one hand, the University of Illinois printed special postal stamps that proudly stated that "Four Colors Suffice" and computer science departments all over

* Or, rather, an unavoidable set of reducible configurations.
† In math jargon, *modulo* means, more or less, "except for."

the country invited Appel and Haken to give talks. On the other hand, many mathematicians wouldn't accept the computer-based proof and were almost offended by Appel and Haken's proposal. Despite the widely accepted idea that all mathematical statements are either provable or refutable, in 1931 an Austrian logician, Kurt Gödel, had shown that no matter how we formalize mathematics, there will always be some statements that can be neither proved nor refuted. Haken suspected that the mathematical community, still shocked by Gödel's incompleteness theorems, wanted at least to believe that theorems with simple statements—such as the four-color problem—should have nice, simple, elegant proofs. Appel and Haken didn't discredit this wish: "Of course, a short proof of the four-color theorem may some day be found, perhaps by one of those bright high school students. It is also conceivable that no such proof is possible." Blostein remembers her father explaining that although the proof seemed inscrutable, it wasn't that complicated—just long. He even considered (I'm not sure how seriously) printing out the configurations, which would likely fill whole rooms with paper, and training a brigade of teenagers to check them by hand. Maybe this would have satisfied the skeptics, but it wouldn't have guaranteed correctness either, as humans are more likely to make mistakes than computers.

Blostein tells me that every Saturday, mathematicians at the University of Illinois would go on a hike,* sit around a campfire, and discuss mathematics. They liked to take bets, and the loser would have to bring a pie to the next hike. One of the bets was proposed by keen hiker Joseph L. Doob, a famous mathematician with remarkable achievements in analysis and probability theory.† He believed that if you opened any mathematical journal on a random page, you'd be able to find a mistake. I'm not sure if he ended up bringing a pie, but his statement wasn't far from the truth—computers reduce the chance of mistakes in proofs rather than create them.

Opinions on Appel and Haken's proof varied to such a degree that mathematicians on opposite sides of the barricade had to rethink what a mathematical proof is. Can we accept a proof that a skilled mathematician cannot

*At first, only men were allowed, apart from the annual day when the invitation was extended to wives and children. As the number of hikers decreased, however, the hikes became more family friendly.

†Analysis deals with approximating mathematical objects with other objects that are easier to handle. One of the branches of analysis is calculus. Probability theory lets us rigorously analyze uncertain events.

understand? What about if a human cannot check the calculations? These questions are more philosophical than mathematical and don't have one correct answer.

Today, computers have become a widely accepted extension of pen and paper, and nobody questions the validity of computer-based calculations. The next step is to accept mathematical proofs done fully by artificial intelligence. It's not the future—it's happening right now! Automatic theorem provers are getting better, and some universities have even started teaching future mathematicians to use them. Personally, in the computer-proof era, I'll miss mathematical elegance, but I'm also excited about all the doors these new tools will open.

Vertices, Edges, and Computers

Mathematicians transform maps into graphs not for aesthetics but to focus on the features pertinent to the problem. Before Euler converted the map of Königsberg into a graph, nobody had thought about counting the number of edges leaving each vertex. While maps encourage us to focus on a specific geographical region, graphs help us to generalize. Thanks to his idea, Euler solved the puzzle not only for Königsberg but for all possible bridge configurations.

By pondering over an apparently insignificant puzzle, Euler started a whole new field—graph theory—which allows us to convert complex real-life problems into a more digestible, visual form. Graphs are arguably even more useful now, in the computer era. Computers don't understand maps, and graphs let us translate maps into clear, numerical inputs. This helps us to solve problems too computationally challenging for even the most skilled person to attempt, from route planning to seating party guests to map coloring. The story of graph theory's development shows us that in math, the lines between theoretical and applicable, simple and hard, and trivial and deep are blurred. Solving even the smallest of problems, like designing a Sunday stroll or coloring a map, can change history.

6

DIVIDED

How to Shape Society

On my first ever parkrun—a weekly timed 5K event organized on Saturday mornings in over 2,000 cities around the world—I got lost. To this day, I have no idea how I managed to get confused on a simple loop around the park. The experience made me doubly in awe of the participants of another 5K that took place in Asheville, North Carolina, in November 2017. Even looking at the map of that route, shown in Figure 6.1, makes me dizzy. It seems that just as one gets into a rhythm, one needs to turn, and turn again, and again. Who would draw such a bizarre route? The answer is short: politicians.

When North Carolina Republicans passed a law forcing transgender people to use restrooms corresponding to the sex identified on their birth certificates, local artist J. P. Kennedy decided he had had enough. He didn't feel that the decisions of the conservative authorities represented the views of the citizens of Asheville, a town so progressive, liberal, and hipster that it has been described as the "Brooklyn of North Carolina." Kennedy's feelings were justified—although the community was overwhelmingly Democrat, both of their representatives in Congress were Republicans with starkly opposing views to most of the town's inhabitants. Somehow, Republicans had managed to divide the voters into two districts and win a seat in both, despite achieving a smaller total number of votes.

It's the meandering boundary of the two congressional districts to which Kennedy wanted to draw the community's attention. So, together with his

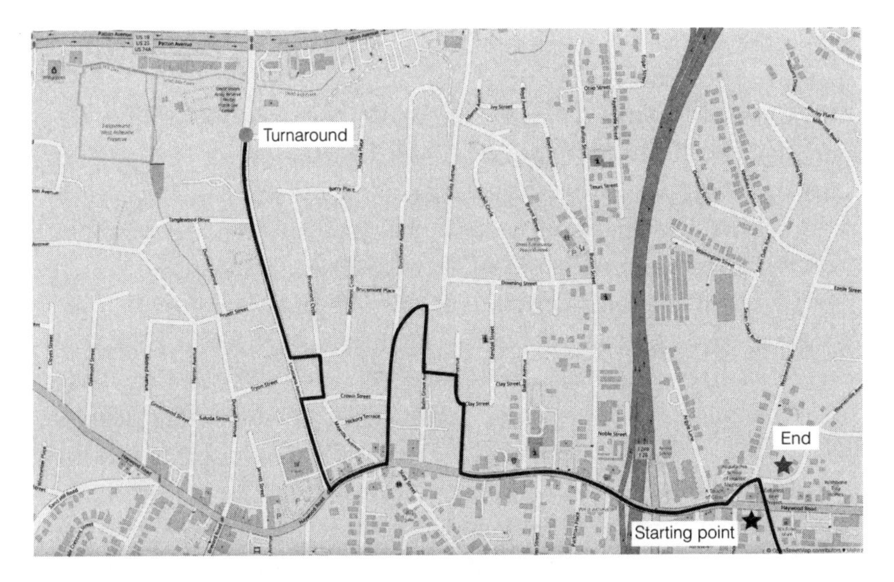

Figure 6.1 The route of Gerrymander 5K followed the wiggly boundary between two voting districts in Asheville, North Carolina.

wife and the local branch of the political grassroots network the League of Women Voters, he organized a race along the curious 5K boundary.

Cracking and Packing

Every ten years, after the federal census, political maps within the United States are redrawn. Voting districts should have approximately equal numbers of voters, but over a decade, people are born, people die, and people move. The new maps should reflect the current state of the population, keeping it as up to date as possible. That sounds fair. The problem is that the map can be divided into equally populated districts in a myriad of ways, some fairer than others.

In the essentially bipartisan United States, the party that gets a higher number of votes in a district—no matter how many—wins the seat. With that in mind, parties have a lot of interest in strategically spreading their voters between districts. To understand that statement better, let's consider Gerrylandia, an imaginary country of twenty-five voters and five districts, each choosing a representative from either the Star Party or the Circle Party. Sixty percent of Gerrylanders support the Star Party, and forty percent support the Circle Party, so it would seem fair if Star Party representatives got three seats and Circle Party representatives got two seats. The simplest option would be to create three districts with five voters each, fully supporting the Star Party, and two districts with five voters each, fully supporting the Circle Party, as shown in Figure 6.2a. In the real world, it's unlikely for a district to fully support one party, but this doesn't mean that it's impossible to achieve the same seat distribution, as Figure 6.2b shows.

The same voters can be divided into districts that would result in very different seat assignments. On one hand, dividing the twenty-five citizens into five districts, each with three Star Party voters and two Circle Party voters, like in Figure 6.2c, would guarantee victory for the Star Party in each of them. On the other hand, the Circle Party, despite being the less popular of the two, could win the majority of seats, like in Figure 6.2d. The latter two maps don't represent the electorate and are likely gerrymandered, which means that district boundaries are intentionally manipulated to favor one group.

Partisan gerrymanderers use different tactics to increase the impact of their voters and decrease the impact of the opponent's voters on the election results. The most important ones are cracking and packing, usually used together on the same map. Cracking involves spreading the opponent's electorate between many districts so that their voice isn't strong enough to

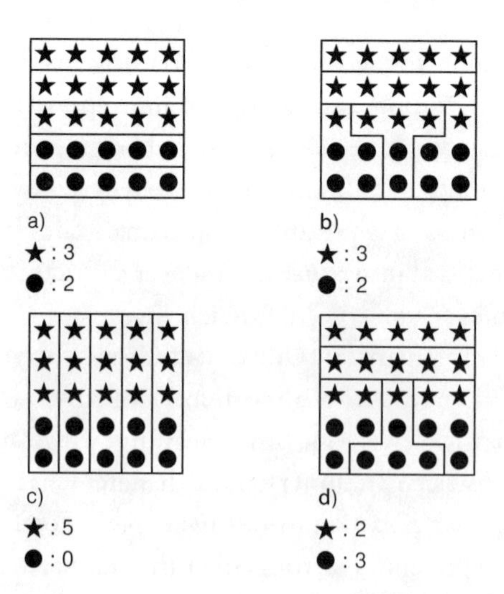

Figure 6.2 Different ways of dividing twenty-five people into five districts give different election results.

win in any of them. For example, in Figure 6.2c, each district has some Circle Party voters, but in each, they're a minority, losing to the Star Party. On the other hand, packing puts as many of the opponent's voters into as few districts as possible. In those districts, the opponent wins with a great margin, but leaves the remaining seats for the incumbent, like in Figure 6.2d. All tactics aim to waste the opponent's votes, that is, prevent the opponent's votes from contributing to the result, which violates the democratic rule that every vote counts the same.

And with that information, we're now ready to understand the bizarre route of Asheville's 5K run, sometimes called the Gerrymander 5K, a race not so much for pleasure but for democracy.

The Secret of Zigzags

Before 2011, North Carolina's Congressional District 11 consisted of the whole liberal town of Asheville and its conservative suburbs. This political mix made it one of the most competitive districts in the American South. It all changed when Republicans took over both houses of the North Carolina Legislature, which put them in charge of drawing new electoral maps after the 2010 census. To make sure Asheville's Democrats wouldn't endanger their seats in the next election, the Republicans split the town between two districts, making this map a textbook example of cracking. Now, Democrat voters did not have enough power to choose their representative in either District 10 or District 11.

Republicans didn't split Asheville at random. With access to detailed voter data, the mapmakers could decide where to draw the line house by house. Complex computer models not only indicate which houses are Democrat and which Republican, they also quantify how likely each person is to vote for a given party. But when a writer and a participant of the Gerrymander 5K named Jeremy Markovich investigated why one house was in District 10 and the house next door in District 11, he came away none the wiser. All houses in the neighborhood looked similar, and so did their residents. He also couldn't find any meaningful differences between the voting preferences of neighboring houses that ended up in separate districts. The mystery was solved by a redistricting expert, Blake Esselstyn.

In 1964, the Supreme Court decided that congressional districts must have almost identical populations; otherwise, some votes would matter more than others. Thanks to detailed voter data and advanced mapping software, politicians can draw districts of exactly equal sizes, down to one person. For example, according to the 2020 census, all fourteen districts in North Carolina have 745,671 voters each, plus or minus one voter. This makes it hard for the court to challenge the district maps—in other words, it makes it easy to get away with gerrymandering. According to Esselstyn, since West Asheville is predominantly Democrat, the mapmakers didn't have to worry about who lived in the houses along the district split, as long as the numbers added up. So, despite Markovich's suspicions, it wasn't a demographic or ideological difference that placed people in separate districts; it was their count. The dense, urban, and rather politically homogenous population of Asheville made it easy for Republicans to gerrymander the district boundary and get away with it. With that, they joined a centuries-old and hugely infamous tradition, practiced across the political spectrum.

The Most Famous Salamander

The name "gerrymandering" comes from the amphibious shape of a state senate district drawn by the then-governor of Massachusetts, Elbridge Gerry, in 1812. His attempt to rig the elections was spotted and criticized in the *Boston Gazette,* which published a cartoon—still famous today, and whose original woodblocks are kept by the Library of Congress—depicting Gerry's district as a salamander: Gerry + salamander = gerrymander.* The same year, the term was reprinted in over eighty newspapers all over the United States.

Although the most famous, Gerry's salamander wasn't the first case of redrawing political maps to benefit one group. In England, after the Magna Carta of 1215, each parliamentary borough could send two representatives to Parliament. As the population shifted, by the eighteenth century some constituencies, known as rotten boroughs, became so small that wealthy landowners could easily influence the voters. On one hand, there was the city of Manchester in the north of the country, whose population skyrocketed during the Industrial Revolution, reaching 100,000 in the early nineteenth century; on the other hand, there was Old Sarum in Wiltshire in the southwest, with a handful of eligible voters. Both these boroughs had

* Initially, gerrymandering was pronounced with a hard *g* (as in *get*), to mimic the pronunciation of "Gerry," but today the pronunciation with a soft *g* (as in *gentle*) dominates.

two representatives in Parliament. Tiny districts such as Old Sarum disappeared only after the 1832 Reform Act, which significantly changed the electoral system of England and Wales. While not gerrymandered per se since they arose from lack of redistricting to reflect changing demographics rather than ill-intentioned redrawing of maps, these unequal boroughs were predecessors of the notorious practice.

In the United States, gerrymandering is at least as old as the country's first constitution. For example, in Virginia's first congressional elections in 1789, Anti-Federalists led by Patrick Henry redrew the electoral map to prevent the Founding Father James Madison from getting a seat in the US House of Representatives (Madison won anyway). Perhaps we should rename the misdeeds of electoral mapmakers "henrymandering."

Today, gerrymandering is a phenomenon almost exclusively linked to the United States. While not entirely immune to gerrymandering, most other democracies have managed to discontinue the practice of redrawing electoral districts to disadvantage an incumbent party's opponents. This is often due to different electoral systems that allow choosing more than one representative per district. In most countries outside the United States, the districts are redrawn by independent commissions rather than the parties themselves, who will naturally struggle to stay unbiased.

This does not mean that in other democracies each vote counts the same. For example, in the United Kingdom, some constituencies are protected by law for geographical reasons. Inhabitants of Orkney and Shetland, Na h-Eileanan an Iar, and the Isle of Wight are guaranteed a representative (two in the case of the Isle of Wight) in Parliament, despite the small populations of these islands. This means that a vote of an inhabitant of the Isle of Wight has a larger impact on the election than the vote of a Londoner—not to benefit a particular party, but to ensure that the interests of the residents of this geographically and culturally unique place are protected.

Attempts at gerrymandering outside of the United States are so rare that they're widely covered in international media, as they often indicate a threat to the country's democracy. For example, the 2017 proposal of the Polish ruling party, Law and Justice, to expand Warsaw's electoral boundaries—making the capital's area larger than the area of Paris or London—would mean that the nearly two million residents of the current Warsaw would get only eighteen representatives while less than one million citizens of the added communities would get thirty-two. Urban Warsaw votes predominantly against Law and Justice, but the small towns and villages that would join the capital have far more Law and Justice supporters. Another example comes from Hungary, where the 2022 elections were largely skewed after the ruling Fidesz party packed the opposition voters into just a few large districts.

While gerrymandering happens all over the world, nowhere is it such an important issue as in the United States. Understanding US electoral problems will help us to recognize when redistricting goes too far, wherever we live.

In Bad Shape

Though gerrymandering is centuries old, today it is more dangerous than ever. Early gerrymanderers could only cause so much harm by redrawing maps by hand, in a hit-or-miss way. Today, politicians have powerful computers and detailed data at their disposal, which allows them to create precise borders that maximize their chance of winning while keeping the map constitutional. The same computers, however, can also be a weapon against gerrymandering. But to identify gerrymandered districts, we first need to define what "fair" means.

When we think about a fair district, we imagine it to have a reasonable shape; it definitely shouldn't resemble an amphibian. In contrast to elongated or weird shapes, we'd like the districts to be compact, resembling Poland rather than Chile. A compact shape encloses an area efficiently, minimizing its perimeter. Among all shapes of the same area, it's always the circle that has the smallest perimeter; in other words, a circular boundary is the most efficient way of enclosing an area.

To find out how much area the boundary encloses, we can find the ratio between the district's area and its perimeter. For example, a one-by-one square has an area of $1 \times 1 = 1$ and a perimeter of $1+1+1+1=4$, which makes a ratio of $1/4$. If we keep the district's square shape but double the lengths of its sides, we get an area of $2 \times 2 = 4$ and a perimeter of $2+2+2+2=8$, which makes a ratio of $4/8 = 1/2$. This makes the measure rather useless because our judgment of the district's shape shouldn't depend on the map's scale.

To improve upon this idea, we instead divide the area by the squared perimeter. Now the ratio depends only on the district's shape and not on its size, and larger ratios indicate more efficient shapes. For example, the ratio for the smaller square becomes $1/4^2 = 1/16$, and for the larger square

$4/8^2 = 1/16$. According to this measure, circles are always more efficient than squares.* This inspired the often-used Polsby–Popper test, which compares the district's area to the area of a circle with the same perimeter.† This works well for theoretical district shapes such as squares, but measuring lengths of real-world boundaries is all but impossible due to the coastline paradox. The length of a boundary depends on the length of the measuring stick, which is not ideal when we're comparing perimeters of different districts, drawn by different people in different years.

Some measures quantifying a district's compactness don't suffer from the same problem, but every shape-based measure is misleading. While it's true that a "weirdly shaped" district is more likely to have been gerrymandered than a nice circle or a square, it's impossible to judge whether a district is fair just by considering its shape. If geometry alone doesn't tell us if a district is fair, what measure can we use?

Who's More Efficient?

In 2015, law professor Nicholas Stephanopoulos and political scientist Eric McGhee suggested focusing on election results rather than district shapes to detect gerrymandering. In any election, each party "wastes" some votes: all votes in losing districts and the votes not necessarily needed to get the majority in winning districts. If both parties waste a similar number of votes, the voters of each party have a similar impact on the results. In gerrymandered elections, on the other hand, one party wastes many more votes than the other. Stephanopoulos and McGhee proposed a tool to measure this imbalance, sometimes used by courts in gerrymandering cases: the *efficiency gap*.

The efficiency gap, as the name indicates, focuses on how efficiently the parties spread their voters between the districts. A party wants to win as many districts as possible by narrow margins, to avoid wasting any votes in already won districts; and in districts it loses, it wants to lose badly, to avoid wasting votes in districts won by the opponent. The efficiency gap is the difference between the votes wasted by two parties, expressed as a percentage

*For a square with side s, the ratio between its area and its squared perimeter is always $s^2/(4s)^2 = 1/16$. For a circle of radius r, this ratio is equal to $(\pi r^2)/(2\pi r)^2 = 1/(4\pi)$. The latter ratio is larger, making the circle more efficient.

†The Polsby–Popper score is given by a ratio between 0 and 1: $4\pi A/p^2$, where A is the district's area, and p is the district's perimeter.

Table 6.1 Example election results proposed by Stephanopoulos and McGhee. To calculate the efficiency gap, we need to find the wasted votes (the sum of lost and excess votes) of both parties.

	Total Votes		*Wasted Votes*	
District	**A**	**B**	**A**	**B**
1	70	30	20	30
2	70	30	20	30
3	70	30	20	30
4	54	46	4	46
5	54	46	4	46
6	54	46	4	46
7	54	46	4	46
8	54	46	4	46
9	35	65	35	15
10	35	65	35	15
Total	**550**	**450**	**150**	**350**

of the total number of votes. To see how it's calculated in practice, let's look at the example from the original paper by Stephanopoulos and McGhee.

Let's imagine elections with ten districts of 100 voters each, whose results are presented in Table 6.1. Party A won districts 1–8, so all votes of Party B in these districts were wasted. Similarly, Party B won districts 9 and 10, so all votes of Party A in these districts were wasted. Since each district has 100 voters, all votes above the required fifty* were wasted; for example, in district 1, Party A wasted $70-50=20$ votes. I'd encourage you to go through the computations yourself, but the table shows that Party A wasted 150 votes, and Party B wasted 350 votes in total. This makes the efficiency gap $(150-350)/1000=-0.2$, or 20 percent in favor of Party A. In other words, Party A won in $20\% \times 10 = 2$ more districts than it would have if both parties had wasted the same number of votes.

In a fair election, both parties waste a similar number of votes, which translates to a small efficiency gap. The authors of the original paper suggest that an efficiency gap larger than 8 percent should trigger a Supreme Court investigation. This clear threshold based on an easy-to-calculate

*Technically, fifty votes would result in a tie, and fifty-one would be needed to get a majority. I'll stick to the 50 percent threshold common in the literature, which doesn't significantly change the results.

Table 6.2 An example of a proportional election detected as "unfair" with the efficiency gap

	Total Votes		*Wasted Votes*	
District	**A**	**B**	**A**	**B**
1	100	0	50	0
2	100	0	50	0
3	100	0	50	0
4	100	0	50	0
5	100	0	50	0
6	100	0	50	0
7	0	100	0	50
8	0	100	0	50
9	0	100	0	50
10	0	100	0	50
Total	**600**	**400**	**300**	**200**

metric sounds like a neat solution to the problem of gerrymandering, but does it really tell us if an election was fair?

Table 6.2 shows an example of another ten-district election with 100 voters each. Party A gained 60 percent of the votes and 60 percent of the seats while Party B gained 40 percent of the votes and 40 percent of the seats, which means that the election was proportional. However, since Party A wasted fifty votes in each of the six districts it won, and Party B wasted fifty votes in each of the four districts it won, the efficiency gap of $(6 \times 50 - 4 \times 50) / 1000 = 0.1 = 10\%$ would be flagged by the Supreme Court.

By its very construction, the efficiency gap penalizes proportionality, which is easier to notice if we express the efficiency gap in a slightly different way. With some algebra, we can rewrite it as the difference between the statewide vote margin, which is the difference between the percentage of votes cast for that party and the percentage of votes cast for its opponent, and *half* the seat margin, which is the difference between the percentage of seats won by the party and the percentage of seats won by its opponent. So, the efficiency gap is zero only if the statewide vote margin is equal to *half* the seat margin, while we deem an election proportional if the statewide vote margin is approximately equal to the seat margin. In our last example, Party A's vote margin of $(600 - 400) / 1000 = 0.2$ was equal to Party A's seat margin of $(6 - 4) / 10 = 0.2$. So, the election was proportional, but the efficiency gap was $0.2 - 0.2 / 2 = 0.1$.

It follows that, according to the efficiency gap, the "fairest" districts are the ones with a 75–25 vote split. There, the winning party wastes twenty-five votes, and so does the losing party, making them both equally efficient. But districts with such a large vote difference between the two parties aren't competitive, which discourages the candidates from listening to either side of the electorate. Why would they bother paying attention to the citizens of districts where the election is essentially won before the campaigning even starts?

The efficiency gap discourages the mapmakers from drawing competitive districts, which is something we should instead be incentivizing. In a close race, a difference of just a few votes in either direction drastically changes the wasted votes tally. Imagine, for example, a district where Party A got fifty-one votes and Party B got forty-nine votes. So, Party A wasted one vote, while Party B wasted all forty-nine votes. But if Party A got two votes less, and Party B got two votes more, the result would flip: now Party A would waste forty-nine votes, while Party B would waste only one vote. A measure so volatile in competitive elections—which, in some way, are the fairest elections out there—cannot be relied upon.

The efficiency gap can indicate if an electoral map is biased, but it can't tell us where the bias comes from. Was the map intentionally skewed toward one party, or is it the effect of the natural distribution of voters? The answer to this question cannot be captured in a single number, but researchers are developing increasingly effective methods of establishing the *intention* behind the electoral map. This complex question requires a holistic approach and a mastery of many fields, from mathematics to political science. To get closer to an answer I was lucky enough to be able to speak to a person who ticks all these boxes.

With a Little Help from the Computer

Wendy K. Tam Cho's email response to my interview request is brief, but it includes a long list of her varied affiliations. According to her signature, she's a political scientist, statistician, mathematician, computer scientist, and Asian American Studies scholar at the University of Illinois Urbana-Champaign. What sounds like a curious amalgam of disciplines is what makes her one of the world's leading experts in gerrymandering detection and prevention. Having studied redistricting for three decades, from so many different points of view, she understands the topic better than anyone else.

I speak with Cho over the phone as she walks around her neighborhood. Experienced in speaking with journalists, she answers my questions precisely and to the point. I can't help asking about the many fields she's working in, which becomes the main theme of our conversation. Far from being empty words on a résumé, her affiliations represent years of studying the fields in depth.

"When you live in a lot of different worlds like that, you realize that they're all connected," she tells me, adding that this realization comes with a thorough understanding of a field. Mathematicians and computer scientists tend to consider political science as an inferior discipline, but Cho stresses that one cannot just read a few news stories and claim to understand politics.

"Every field is very deep and there's a lot to understand," she says.

Cho, along with other mathematicians dedicated to fighting the increasing threat of gerrymandering, recognizes the potential of powerful computers to detect gerrymandered electoral maps. With clever algorithms, she generates many possible district maps of a state in question, which, by definition, aren't intentionally biased. When she compares the resulting large set of artificial maps with the electoral map proposed by politicians, she can see whether the latter disproportionally favors one of the parties.

The artificial maps must fulfill some legal criteria. For example, each district must contain an approximately equal number of voters; it must be contiguous, that is, one must be able to travel between any two places within the district without crossing into a neighboring district; and it must respect political boundaries, such as town or county lines, whenever possible. Cho makes sure that the computer-generated maps satisfy these legal criteria, so that she can compare them to the existing map.*

To generate the maps, Cho doesn't use any partisan data, so the artificial maps aren't gerrymandered. This doesn't mean, however, that the maps she's working with aren't biased toward one of the political parties, since they're reflecting real electorates that tend to cluster together, creating urban–rural political divides. The question Cho is trying to answer isn't whether a given map is unbiased, but whether it's more biased than the population characteristics would predict—in other words, if the map designer had less than honest intentions. If the actual map is more biased toward the incumbent

* The criteria vary between states, but most states require at least the basic criteria listed here.

party than most of these algorithm-generated maps, it is likely to have been gerrymandered. If, on the other hand, the actual map favors one party to an extent similar to the artificial maps, then there's no evidence that it was created with partisan gain in mind. Cho's idea cannot *prove* that an electoral map has been gerrymandered, but it can give a strong indication either way.

The difficulty lies in developing an efficient method of generating millions of artificial maps, which is a task so computationally demanding that it requires the world's fastest supercomputers. A state can be divided into districts in a myriad of ways; even a slight change to a boundary creates a new artificial map. Districting boils down to distributing little regions, called precincts, between larger districts. Figure 6.3, for example, shows all ten possible divisions of a nine-precinct square into three-precinct contiguous districts. The number of possible maps grows quickly as we increase the number of precincts: a four-by-four square can be divided into contiguous four-precinct districts in 117 ways, a five-by-five square can be divided into contiguous five-precinct districts in 4,006 ways, and a nine-by-nine square can be divided into contiguous five-precinct districts in over 700 trillion ways, which is more than the number of cells in a human body. We don't even know how many possible divisions of a ten-by-ten square there are, so generating all possible electoral maps of California, with over 20,000 precincts, or even Alaska with a modest 441 precincts, is way beyond our reach.

Because no computer can generate all possible districting maps even for the smallest of states, scientists build algorithms to sample just a fraction of them. Not just any fraction though, but a set of maps representative of

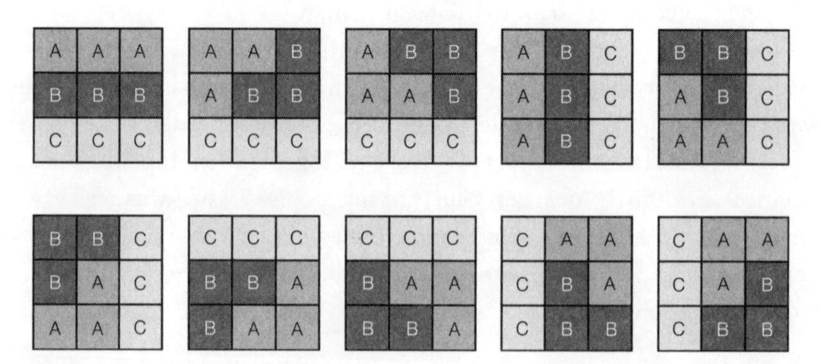

Figure 6.3 A three-by-three square can be divided into three-precinct districts in ten different ways.

all the possible divisions of a state into electoral districts. The simplest algorithm to generate electoral maps is called "random seed and grow." It starts by randomly choosing a precinct, which will become a "seed" of the first district. Then, it lets the seed "grow" by attaching to it random adjacent precincts until the total number of voters in the resulting blob gets close to the target. Now that the first district is ready, the algorithm plants another seed, which will become the second district, and so on, until all precincts are assigned to electoral districts. Albeit quick and simple, this method isn't reliable, as experiments have shown that the generated set of maps can be randomly biased toward one party, deeming any comparisons with the real map misleading. Aware of this problem, researchers have been working on developing alternative techniques.

Cho and colleagues, for example, developed a so-called evolutionary algorithm, which mimics the changes that happen in populations of animals or plants over time. In this case, instead, it's the maps that "mutate" and "mate," creating offspring. First, the computer uses "random seed and grow" to generate a few hundred initial maps, which are ranked according to a predefined quality criterion. Some maps "mutate," which means that randomly selected precincts are moved to adjacent districts. Two maps of high quality—to mimic the survival of the fittest—"mate," which involves overlapping them and merging some of the "children" districts to obtain a new map. The worst maps "die," that is, they are removed from the final set. And so, the population evolves, generation by generation, until the sample is large enough.

Evolutionary algorithms aren't the only option. In 2014, a team of scientists at Princeton University generated electoral maps using a popular sampling method called Markov chain Monte Carlo (MCMC). To visualize how MCMC works, imagine that all possible maps are scattered across a hilly terrain. The "better" a map is, according to a predefined criterion, the higher it will be placed. MCMC walks around the area, spending more time in places higher above the ground, and so collecting more plausible maps. At every step, the algorithm picks a random location, and if it's higher than the current location, it moves there—otherwise, it stays put. After it finishes, we'll have a list of locations it has visited and the corresponding maps, with the high-quality maps placed higher above the ground appearing on the list more often.

Like with the four-color theorem, the key to applying MCMC to this problem is turning maps into graphs. This not only lets us communicate the problem to a computer but also offers the techniques from graph theory that let us prove that the method—under some conditions—will lead to a representative sample of maps. One of these conditions is that the algorithm must run for a sufficiently long time, but it's hard to say exactly what this minimum time is. That's why it is important to find ways for the algorithm to explore our hilly space quickly and efficiently.

A major improvement came in 2019 from the MGGG Redistricting Lab, a research group at Tisch College of Tufts University led by mathematician Moon Duchin. Most MCMC algorithms start by generating an initial map, for example with "random seed and grow," from which the algorithm will begin exploring the hilly terrain. It's the way it decides where to go next which distinguishes variations of the algorithm. In the original 2014 algorithm, sometimes called "flip MCMC," each step involves swapping two randomly chosen precincts from two adjacent districts. With this approach, individual steps are fast, but the algorithm explores the hilly terrain very slowly as each map barely differs from the next one, so they likely lie close to each other. The algorithm proposed by Duchin and colleagues, called ReCom, at each step instead picks two random adjacent districts, merges them, and randomly redistributes precincts between these two districts. This requires more changes during a single step, which makes steps slower than in flip MCMC. Each ReCom step, however, produces a map that differs much more from the previous one than each step of flip MCMC. This means that with ReCom, we're more likely to obtain a representative sample much faster than in previous MCMC versions.

As I write these words, multiple groups of mathematicians are developing new methods of efficiently sampling millions of possible maps, which help establish whether the actual electoral map has been gerrymandered. For example, when Wendy Cho and colleagues applied their algorithm to the 2011 map of Maryland congressional districts, which had been challenged by a bipartisan group of voters, it generated one quarter of a billion legal maps, almost all favoring Democrats. The original map, however, was biased more strongly toward the Democrats than 99.79 percent of all generated maps, which would be very unlikely if the Maryland map hadn't been gerrymandered.*

* Maryland's 2011 electoral maps were challenged in court and the *Lamone v. Benisek* case went on for many years. Finally, in June 2019, the Supreme Court issued a ruling. It did not rule on whether the Maryland map had been gerrymandered; instead, it decided that the federal courts, including the Supreme Court itself, cannot judge gerrymandering cases. This essentially means that electoral maps are fully in the hands of the states.

If there exists such a powerful tool to detect gerrymandered maps, why does the United States still have so many unfair districts? Cho believes that while algorithms can do a lot to illuminate the problem, they're useless if the legal systems don't adapt to the idea of fighting gerrymandering with technology. Understanding Cho's method requires deep technical knowledge, which the courts don't have, and the law cannot be created based on tools that the legislators don't understand. Implementing algorithms in the fight against gerrymandering requires acceptance and adaptation from the legal system and wider society. That's why it takes an expert in both a quantitative field and politics, such as Cho, to create an essential connection between these two worlds. Cho goes as far as to say that "otherwise, it [the algorithm] is not going to work; it's going to create worse outcomes than if we didn't have it at all." She adds that solving gerrymandering isn't a matter of technology, but "it's a human question because humans are capable of drawing fair maps."

A human judgment is needed even when an algorithm leaves no doubt that a map has been gerrymandered.

Good Earmuffs, Bad Earmuffs

In some circumstances, a gerrymandered map can do more good than harm and no algorithm can decide if that's the case. Looking at Illinois's Congressional District 4, it's easy to see why Republicans accused Democratic designers of this mostly Democratic district of gerrymandering. The Earmuff District, as it's widely known because of its shape, is in some places only about 150 feet wide.* Figure 6.4 shows that its western part consists only of a highway segment, excluding residential areas on both sides of the road. Democrats openly admit that this bizarre configuration was indeed gerrymandered—but for a good reason. What looks like yet another example of outrageous manipulation of elections to favor one party over another turns out to be a legal attempt to give voice to often marginalized communities.

The Voting Rights Act of 1965, signed by President Lyndon B. Johnson, was intended to stop racial and ethnic discrimination in voting and was later extended to include language minorities. It obliges mapmakers to

*In January 2023, the boundaries of Illinois's Congressional District 4 were redrawn. For illustration purposes, I'll stick to the previous, earmuff version of this district.

Figure 6.4 Illinois's Congressional District 4 looks like a pair of earmuffs formed by two neighborhoods connected by a highway.

ensure that minority voters can choose their representatives. Remember that each district chooses just one representative, so even a relatively large minority—making up, let's say, 40 percent of the total district's population—will never get to have someone representing their interests. One way around it would be to change the system, so that rather than "winner takes all" it resembles the proportional representation used in dozens of countries from Albania to Uruguay: votes from all over the country get aggregated, so a 40 percent minority would get about 40 percent of the seats. This, however, would go against an important American value of local representation. Instead, the Voting Rights Act makes sure that some districts are drawn so that the minority group becomes a majority within the district.

As with many redistricting issues, there's no hard and fast rule as to what such majority–minority districts should look like, but the Supreme Court has established three factors that must be satisfied. First, the minority group must be large enough and compact enough to create a reasonable district where it would become a majority—for example, Black voters scattered all over the state shouldn't form a separate district. Second, the minority group must be politically cohesive, that is, most group members must consistently prefer similar candidates. Third, the majority in the current district has historically avoided choosing a representative belonging to the minority—for example, if the white majority never picks a Hispanic candidate put forward by the Hispanic community. If these three rules are satisfied, the map should include a district where the minority voters have a real chance of winning the elections.

With this in mind, let's look again at the Earmuff District. The notorious highway connects two Hispanic and Latino communities in the Chicago metro area, leaving a gap in the middle for another minority district—predominantly Black Congressional District 7. This way, both Hispanic and Black voters can choose a candidate who will best represent their interests.

Nicholas Stephanopoulos is such a fan of the Earmuff District that he had this shape printed on his and his wife's wedding cake. To him, it proves that "not all funnily shaped districts are bad," and sometimes we need a rather ridiculous boundary to protect the rights of people who have been silenced for too long.

All Votes Count

When Wilfred Jones's sister suffered a heart attack, it took the ambulance over an hour to arrive. Jones realized that many of his neighbors living in the southern part of Utah's San Juan County, a Navajo reservation, had also lost loved ones because paramedics had to travel too far to get to emergencies. Deciding that enough was enough, Jones joined the board of the Utah Navajo Health System and started lobbying with the county commission for an ambulance station closer to the reservation. During each vote, one commissioner supported Jones's appeals while the other two didn't. Fed up with years of unsuccessful lobbying, the Utah Navajo Health System decided to use tribal funds to buy ambulances, build a garage to house them, and train volunteers.

The Navajo had always been lacking basic services, from appropriate health care to electricity, running water, and roads. Whatever they asked the county commission for, two out of three commissioners would reject their appeal—all because of extreme gerrymandering.

Utah granted Native Americans voting rights only in 1957 but, even after the passing of that momentous legislation, Native voters were denied registration, and non-English speakers and the illiterate were unable to participate in English-only elections. In San Juan County, the commissioners were elected at-large, which means that all eligible voters would choose their favorite candidates, and the three top choices would get the seats. The county didn't get its first Native commissioner until 1986, when the US Department of Justice, invoking the Voting Rights Act, forced the county to create three one-member voting districts. While two districts remained predominantly white, the third, southernmost one gave the Navajo the majority needed to elect their first true representative.

This, however, didn't solve the problem of the Voting Rights Act violation. While in District 3, known as the Indian District, the Navajo had an overwhelming majority, in the other two districts, fewer than one-third of the voters were Native Americans. They had a chance to win only one seat out of three, although over half of the county's population were Navajo—in other words, the mapmakers packed Navajo into District 3 to avoid them having an impact on the voting outcome in Districts 1 and 2.* To make things worse, while counties are obliged to redraw their maps every ten years to account for population changes, these district lines didn't change for decades.

In 2012, Wilford Jones and other outraged Navajos sued to redraw the maps to comply with the Voting Rights Act. Six years later, after multiple appeals rejected by the white commissioners, the county was finally given a more balanced district map: predominantly white District 1, mostly Navajo District 3, and a competitive District 2 with two-thirds of the population being Native Americans. In the elections that followed, for the first time in the county's 140-year history, the Native Americans got the majority vote in the county commission.†

The newly elected commissioners had a formidable challenge ahead. In 2016, Barack Obama turned the sacred Navajo site Bears Ears into a na-

*The same concerned school districts: Navajo could choose only two out of five school board members.

†Like many residents of the reservation, District 2's Navajo candidate Willie Grayeyes would pick up his post from a post office box in Arizona due to the relatively short distance and the good quality of postal service. This led to yet another attempt to disqualify the competition: The San Juan County Clerk removed him from the ballot due to his alleged lack of residence in the county. The judge restored Grayeyes's candidacy, and at the time of writing, Grayeyes holds the commissioner's seat.

tional monument but, a year later, Donald Trump reduced its area by 85 percent to allow drilling in this oil-rich region. Since then, this part of San Juan County has become a battlefield between the Republican leaders supporting Trump's decision and the local community trying to protect their heritage. Thanks to the power shift in San Juan County, however, the commission officially supported the interests of the Navajo people. After years of intense campaigning, in 2021 they were able to celebrate when Joe Biden restored the original area to national monument status.

The new maps have not provided immediate change to the Navajos' living conditions; it's still one of the poorest communities in the country, struggling to satisfy residents' basic needs. The dirt roads are still covered in snow for days before the snowplow arrives. Navajo children spend hours on buses to get to the closest school, and many households don't have access to clean running water. But now, at least, Navajos have someone to speak up for them.

Change Is Possible

Gerrymandering might seem like a map-drawing exercise interesting only to politicians and mathematicians, but we've seen that it impacts us all. Fair maps give voice not only to the rich and powerful but also to the less privileged. Ill-intentioned gerrymandered maps, on the other hand, create a system where the needs of the most vulnerable are ignored. We've explored various ways of detecting gerrymandered maps—but is it possible to prevent the practice in the first place?

When redistricting is done by the incumbent party, the mapmaker has a vested interest in gerrymandering the districts so that their party stays in power. This could be avoided if the task of redrawing the electoral maps fell to an independent, nonpartisan commission. A few US states—for example, California, Colorado, and Washington—have already done this. Redistricting taken out of the hands of politicians ensures that any bias in resulting maps isn't politically motivated, but it doesn't eliminate bias. This is because voters themselves are making partisan gerrymandering easier.

Try as you might, you can't gerrymander people who tend to live together by choice. By opting to live close to like-minded individuals, voters have tended to separate themselves into Democratic cities and Republican suburbs and rural areas. Then, because Republicans are more spread out, they usually waste fewer votes and get more than their proportional share of seats.

Some political scientists suggest a radical solution: if more Democrats moved out of the cities, the distribution of seats would become more balanced. We can't realistically tell people where they should live (although we've tried, as the next part of the chapter will show), so we need an alternative solution.

Gerrymandering happens because each voting district chooses only one representative. In multimember districts, common in European countries, more than one candidate wins a seat. This way, if an eight-district state became one big eight-member district, the top eight candidates would get seats, no matter which part of the state they came from. This would better represent the true preferences of all citizens of the state, and maybe even more importantly, it would encourage the candidates to fight for every vote, without excluding voters outside their small constituency.

Since 1967, Congressional multimember districts have been illegal to prevent states (mostly in the South) from drawing large districts to dilute Black votes. The suppression of minority groups remains a danger, so the system of counting votes would need to change to ensure that marginalized voices are heard. In the ranked-choice voting system, for example, voters rank their candidates in the order of preference, which eliminates the least-preferred candidates. Studies suggest that multimember districts with ranked-choice voting contribute to proportional representation of minority groups.

Implementing these changes—from independent redistricting commissions to alternative voting systems—wouldn't be easy and, by itself, wouldn't guarantee fair elections. But, as other democracies show, the change is possible and worth aiming for.

Still Segregated

Electoral districts aren't the only example of maps being manipulated to benefit certain sections of society at the expense of others. Gerrymandered school districts are similarly controversial because they can determine the present and the future of our youngest citizens, impacting educational advantages or disadvantages that shape individual lives and whole societies. Parents are willing to go to great lengths to give their children the best possible—in their opinion—education. That's why, in 2012, residents of Gardendale, Alabama, started campaigning to create a new school district, separate from Jefferson County School District, where their children were currently assigned.

In the United States, school districts have all but total power over primary and secondary public education, from assigning children to schools to allocating resources and designing curricula. There is little overarching government control at the national level. As opposed to voting districts, there aren't many rules surrounding school districts. No wonder, then, that properties in highly ranked school districts cost more than in otherwise similar neighborhoods, and many parents (and future parents) are willing to pay extra for a house close to a renowned school. As not all families can afford to move into a more expensive zone, the country's already vastly unequal access to education is reinforced.

Brown v. Board of Education, a groundbreaking 1954 decision by the US Supreme Court, deemed racial segregation in public schools unconstitutional. The case sprang from the efforts of the National Association for the Advancement of Colored People (NAACP) to challenge the segregated school system in the late 1940s. When Oliver Brown's daughter was refused entry to an all-white school near their house, he filed a lawsuit, which led to the historical success. Brown's win, however, did not mean that American schools magically became integrated; they're as segregated today as they were then. Over half of all school-age children are enrolled in racially segregated districts, where more than three-quarters of students are either white or nonwhite.

A significant part of school funding comes from local taxes, which means that wealthier communities create wealthier schools. In the United States, race and wealth are deeply interlinked, making predominantly white schools much better off than nonwhite. On average, nonwhite school districts get $2,226 a year less per student than white districts, which adds up to a $23 billion gap.*

Money was one of the reasons cited by Gardendale's parents for school district secession. Referring to children commuting to their shiny and well-resourced new high school, one of the parents wrote on Facebook: "Those [nonresident] students do not contribute financially. They consume the resources of our schools, our teachers and our resident students, then go

*This is the difference between the total cost-adjusted state and local revenue of white and nonwhite districts: about $153 billion and $130 billion, respectively.

home." In other words, Gardendale residents weren't willing to share their resources with students from poorer, mostly nonwhite communities. In Jefferson County School District, one in five children lives in poverty, and most pupils are nonwhite. In Gardendale as a separate district, both the poverty rate and the proportion of nonwhite students would halve. This would create a new district for the white, well-off children of petitioning parents and reduce the opportunities for the less privileged students left behind.

School districts are drawn by school boards, whose members are elected by local communities. Like creators of voting districts, these people often have their own interests in mind, but this type of gerrymandering isn't much spoken about. Yet, school district secessions are quite common. Since 2000, dozens of neighborhoods have left the district to which they had originally been assigned. In Alabama, for example, any city with at least five thousand residents can create a separate school district. In Jefferson County, so many weirdly shaped school districts have already seceded that the current map would make the most gerrymandered electoral district map look innocent. But in this instance, the county school system had had enough, and they sued Gardendale. The court pointed out that the racially motivated split would violate the desegregation order, thus forbidding the secession. However, the court's decision was more an exception than the rule, and due to gerrymandered school districts, nonwhite students continue to be discriminated against all over the United States.

Segregated Schools in Segregated Communities

How does a student end up in one public school and not in another? The system of school assignment varies from country to country but, in many, it's based on a school zoning map. In the United States, school-age children living in a school's so-called attendance zone (which in some US states, and in the United Kingdom, is called a catchment area) should, in theory, enroll there. But attendance zones don't necessarily assign children to their closest school because, over the years, many of these zones have been gerrymandered to separate white, wealthy households from Black and less well-off ones.

Assigning children to schools closest to their homes might not be the best idea either. For decades, the federal government worked hard to create segregated neighborhoods. In the 1940s, they started supporting home loans for white, but not for Black families, which created de facto ghettoes within cities. To make things worse, the government also backed development loans in the suburbs that didn't allow Black people to move in, which created predominantly white residential areas. So, after *Brown v. Board of Education,* many white families simply moved out of the cities to avoid sending their children to newly integrated schools. Over the years, suburban schools got a reputation as "good" and city schools as "bad," regardless of their students' potential.

Some cities, realizing how grave the situation had become for students of color, introduced busing, which involved sending school buses to drive children to schools outside their attendance zones. In many US cities, this desegregation attempt wasn't particularly successful, as white families would opt to move away, send their children to private, fee-paying schools, or protest and threaten the policymakers until they gave up on the integration efforts. And, as we've seen, they could also create separate school districts. This desegregation policy has been tried outside the United States. In 2006, the Danish city Aarhus started busing children from schools where many students spoke Danish as a second language to schools where most students were native Danish speakers. Unfortunately, it seems that this strategy worsened both the academic results and well-being of these children.

Researchers have found that the socioeconomic and ethnic composition of the neighborhood largely impacts school segregation in countries as diverse as Finland, Chile, and Australia. Interestingly, this holds even in countries where the home address has little or no impact over the school choice. For example, in Amsterdam parents can enroll their child at any primary school they wish—with priority in the eight most local schools—and still, they tend to select the closest institution. This perpetuates the division into "black" schools, usually defined as schools where over 60 percent of pupils belong to ethnic or cultural minorities (regardless of their skin color), and "white" schools, where minority students make up less than 30 percent of the student community. It's unfortunately not surprising that "white" schools tend to have better resources—which often translates to higher academic achievements—than "black" schools.

We can't deny that segregated schools harm children worldwide. But how can we identify the problematic schools?

Measuring Segregation

Tomás Monarrez, a researcher at the Urban Institute, a think tank in Washington, DC, has devoted his career to studying racial and socioeconomic equity. When I meet him over Zoom, it is easy to forget that this young, chatty economist in a baseball hat is a leading expert in school segregation. After exchanging opinions on the previous night's football World Cup game (our respective teams, Mexico and Poland, drew, so there is no animosity between us), we discuss what prompted him to get interested in school boundaries. He tells me that "there was more theory than data" in the field, so, being an empirical researcher, he took a stab at the topic.

The first challenge was finding a way to measure school segregation. People tend to talk about "segregated" and "integrated" schools, but these concepts are relative to the underlying school system. A mostly white school in a mostly white neighborhood is expected; a mostly white school in a diverse neighborhood is more problematic. Let's take a middle school in Milwaukee, Wisconsin, where 36 percent of students are Black or Hispanic.* If we simply look at this statistic, this school doesn't seem particularly segregated; there is a considerable nonwhite student population. If we zoom out, however, we'll see that 80 percent of children in the Milwaukee School District are Black or Hispanic, so this school is disproportionately white. But how segregated is it with respect to neighboring schools?

To assess how equal (or unequal) a given school is, with the whole school district as a baseline,† Monarrez and colleagues developed the Segregation Contribution Index (SCI). SCI measures how much an individual school contributes to the segregation within a district, putting the school's racial imbalance in a broader context. More precisely, it assesses what would happen to the whole district's segregation level if the school's racial composition exactly reflected the district's composition. This allows the policymakers to identify the schools that contribute most to the system's segregation and whose integration might bring the largest benefits to local students.

* The authors group students from these two different backgrounds because, while they face individual challenges, both Black and Hispanic students consistently get lower academic results than white and Asian students.

† The researchers considered three different baselines: school districts, counties, and metropolitan areas. To keep things simple, I'll stick to school districts.

In the case of Milwaukee, a school with 80 percent Black and Hispanic students wouldn't contribute to the district's segregation; it would simply reflect the underlying demographics. Such a school would have an SCI of zero. A school with a lower or higher share of Black and Hispanic students would have a nonzero SCI—the further from 80 percent its share of Black and Hispanic students is, and the larger the school, the more it contributes to the total segregation.

After painstakingly assigning SCIs to almost all schools in the United States—over 100,000 in all—Monarrez and colleagues concluded that most schools' segregation levels do resemble their district's racial composition. One in three schools, however, is the culprit of more than 10 percent of the school segregation within the district. This means that if these schools exactly reflected the district's racial composition, the district's segregation would decrease by more than 10 percent. In other words, such schools are even more segregated than their local communities.

Problematic Lines

If school segregation was merely a consequence of residential segregation, a school in a diverse community would have a diverse student body—but that's not always the case. Around the globe, residential segregation only partially explains school segregation. In Amsterdam, "black" and "white" schools appear even in ethnically mixed communities. Given a choice between two local schools, most parents opt for a school where most pupils "look like" their child—and thus, "black" schools become "blacker," and "white" schools become "whiter." In mixed neighborhoods, it's the parental choices and insufficient integration policies rather than the socioeconomic map that explain school segregation. As countries have been increasingly taking the power of school choice from the authorities and giving it to the parents, school segregation levels seem to have worsened.

In places where parents don't directly choose the school, they often still contribute to segregation by influencing attendance zone boundaries. In the United States, after *Brown v. Board of Education,* while some families fled to the suburbs to avoid sending children to newly desegregated schools, other white families remained in the cities and lobbied for attendance zones that would separate their communities from neighboring Black ones. Sometimes this was as simple as drawing an east–west border as opposed to a north–south one, as happened in Detroit, where white families tend to live north, and families of color south of the original boundary. This created drastically different racial and ethnic compositions of neighboring attendance zones.

Monarrez and his team looked at tens of thousands of attendance zone boundaries all over the United States to find the most segregating ones. Sometimes, they found that one side of the boundary is mostly Black or Hispanic, and the other side—we're talking about a mere 500 meters away!—is mostly white. This means that while one school might have highly skilled teachers, well-equipped computer labs, and efficient air-conditioning, a school a short walk away might have none of these resources. It's possible for kids who live next to one another—not just in the same city, but often on the same street—to have unequal educational opportunities from the beginning of their lives. The researchers have no doubt that this is racial gerrymandering in a new context. Residential segregation heavily impacts school segregation—it's hard to create diverse classrooms in a racially homogenous neighborhood—but it doesn't tell the full story.

Monarrez's team found over two thousand pairs of schools whose neighborhoods, although separated only by the attendance boundary, have vastly different racial compositions. They wanted to find out to what extent these boundaries ameliorate or aggravate these differences. The results varied: most schools reflected the underlying segregation, some schools made it worse, and some made it better. On average, US public schools are slightly less segregated than neighborhoods, but the potential of attendance boundaries to integrate children is still unfulfilled. Monarrez tells me that "it's not as optimistic as it could be, but [. . .] when I started the project [. . .] I thought I was going to find very racist boundaries everywhere." Instead, it turned out that some districts are actively redrawing the maps to integrate children from different backgrounds, gerrymandering to promote integration instead of segregation.

How to Close the Gap

While racially motivated attendance zone boundaries are disheartening, sometimes the inequalities at play can be improved by just a slight adjustment of the line. This redrawing of the lines is not going to magically solve the problem of racial injustice, which runs deep through society, but even small improvements can have a massive impact on children's lives. And, even more importantly, more inclusive boundaries would send a clear message that at least some citizens don't accept the legacy of discriminatory practices.

The possibility of adjusting catchment areas has also been studied outside the United States. For example, researchers developed an algorithm to optimize catchment areas in Swiss cities. First, they divided the cities into neighborhoods, each containing areas closest to one school. They assessed how each neighborhood scored according to the concentration index, which they defined as an average of two values: the proportion of pupils from foreign-language families and the proportion of pupils whose parents have a low educational background. Concentration indices differed significantly between neighborhoods, and the researchers aimed to adjust catchment areas to minimize these differences. In their research paper, they described their algorithm as a board game, in which schools swap small parts of their catchment areas, trying to get their concentration index as close to the city average as possible. The algorithm showed that even small adjustments to school zoning maps could lead to more diverse schools. Importantly, these changes would neither significantly extend the students' commutes nor demand increased capacities of schools. Sometimes, not much is needed to give all children a fair—or, at least, fairer—start.

Children will face unequal opportunities due to their race, differences in their family's wealth, and their home addresses. Although public school systems have the power to decrease this gap, so far, because they are all too easy to manipulate, they have largely failed millions of young citizens. But this can change, and possible solutions go beyond redrawing the maps.

EdBuild, a limited-term organization set up in the United States to bring greater fairness to the way states fund public schools, proposed concrete steps that could give seven in ten American students equal or increased funding without any raises in taxes or debt. States are trying to match the local-tax-based funding in low-wealth districts, but states' revenues come from less reliable sources than property taxes, such as income and sales taxes. To solve this problem, EdBuild suggested splitting the two functions of school district borders. Their first role is administrative: they assign children to schools and areas to school boards. The second role is financial: they specify taxes from which properties fund local schools. We've seen that, ideally, we should redraw the borders corresponding to the first function to promote equality and integration. The problem is that such proposals tend to receive a lot of backlash from white, wealthy families looking to protect their own interests. Alternatively, we could keep the current administrative borders and expand tax borders. This way, poorer schools would benefit from property taxes in wealthier neighboring areas. By doing this, EdBuild estimated that around three-quarters of low-income students would receive equal or greater funding, and the $23 billion funding gap would decrease to $9.5 billion. It's a simple and almost cost-free solution which could improve the lives of millions of vulnerable children, without significantly disadvantaging the more fortunate ones.

Researchers from a European Commission-supported project, European Cities Against School Segregation (ECASS), provide other advice on fighting school segregation. Besides defining socially heterogeneous catchment areas, they suggest introducing district-wide quotas for socially disadvantaged children. For example, a district might decide that each year, 15 percent of all students admitted to each school must come from disadvantaged backgrounds. Of course, such structural changes need time to create a meaningful difference, so in the meantime ECASS advocates for providing segregated schools with additional resources, from funding to experienced teachers. This would help to level out the quality of education children receive, regardless of the school they attend.

School segregation is a large and widespread issue, impacting young people all over the world. Recognizing the problem is the first step to desegregating schools; now it's time to bring in solutions to create equal opportunities for every child.

By shaping policy decisions, maps have real power over our lives. Gerrymandering—whether of voting districts or school attendance zones—leads to unequal distribution of resources, increasing what are already large ethnic and socioeconomic inequalities. A single, abstract line might give two neighbors who find themselves on either side access to different schools, health care and funding, thus changing their life opportunities. Maps can be redrawn to unite instead of divide and to aid instead of hinder. It's for us to decide.

7

FOUND

How to Save a Life

Overcrowded, dirty and smelly, nineteenth-century London wasn't a particularly nice place to be—at least not for humans. For germs, on the other hand, it was a paradise. In 1831, cholera arrived in England. Tens of thousands succumbed to this nasty disease, which causes vomiting, diarrhea, and, in turn, dehydration so severe that it leads to death, often within hours. As with other contagious diseases, cholera was attributed to bad air, which was commonly called "miasma." Doctors thought that people caught the disease by inhaling the "cholera mist," which seemed a logical proposition given that cholera often appeared in places that smelled terrible. Places like nineteenth-century London.

A young doctor by the name of John Snow first encountered cholera as a nineteen-year-old apprentice sent to treat the sick in a mining village close to Newcastle. There he meticulously documented his stay and started questioning the theory of miasma. Why, he asked, was it the coal miners who would mostly get sick, not people who worked or lived close to the smelly sewage dump? He also wondered how a disease that started in the intestines, not in the lungs or the throat, could be caused by inhaling bad air instead of ingesting poison. This line of reasoning wasn't entirely correct, but he was on to something.

By the time of London's third cholera outbreak in 1854, Snow had moved to the capital and had become an established physician. His reputation as an anesthetist was so solid that he had been asked to anesthetize Queen Victoria during her royal labors. In addition to developing modern anesthesia,

Snow continued to study modes of transmission of cholera. In 1849, in his pamphlet *On the Mode of Communication of Cholera,* he argued that cholera is most likely spread through water contaminated with the bodily fluids of an infected person, although he admitted that his hypothesis needed more evidence. The examples he described in the book, however, were compelling.

In a brief period when London was cholera-free, Snow studied the areas that had suffered the most during the previous outbreaks. In one case, he looked at two rows of houses on opposite sides of the same alley. On one side, nobody had died of cholera; on the other side, the disease had taken dozens of victims. If bad air was the culprit, surely a narrow passage wouldn't be a strong enough barrier to protect half of the inhabitants? A further investigation showed that on one side, the waste was flowing toward the well that provided the residents with drinking water, while on the other side, the waste was flowing away from the well. Snow realized it was the residents who drank contaminated water who got sick, and only they. Though he described it all in his pamphlet, it didn't persuade the miasma believers—more evidence had to follow.

A Natural Experiment

Cholera didn't discriminate: rich and poor, men and women, children and the elderly would suffer from the disease and quickly die. But why did some people contract it while others avoided it? Snow set out to find a pattern. After poring over municipal records, he noticed that the areas of London that got water from different companies had vastly different cholera death rates. Southwark & Vauxhall (S&V) and Lambeth water companies brought water from the same River Thames, so why would cholera kill 315 out of 10,000 residents of the areas supplied by the former, and only 37 by the latter?

The Thames was not only the source of running water for Londoners but also the endpoint for the London sewer system. While sewage still ends up in the river today, it's thoroughly treated beforehand, which makes it safe for the environment. This wasn't the case in the mid-nineteenth century: raw waste would flow straight into the river, carrying all the germs it had encountered along the way.* S&V carried this water "in a most impure

*A few years later, in July and August 1858, all the untreated waste made the Thames smell so badly that this period became infamous as the Great Stink.

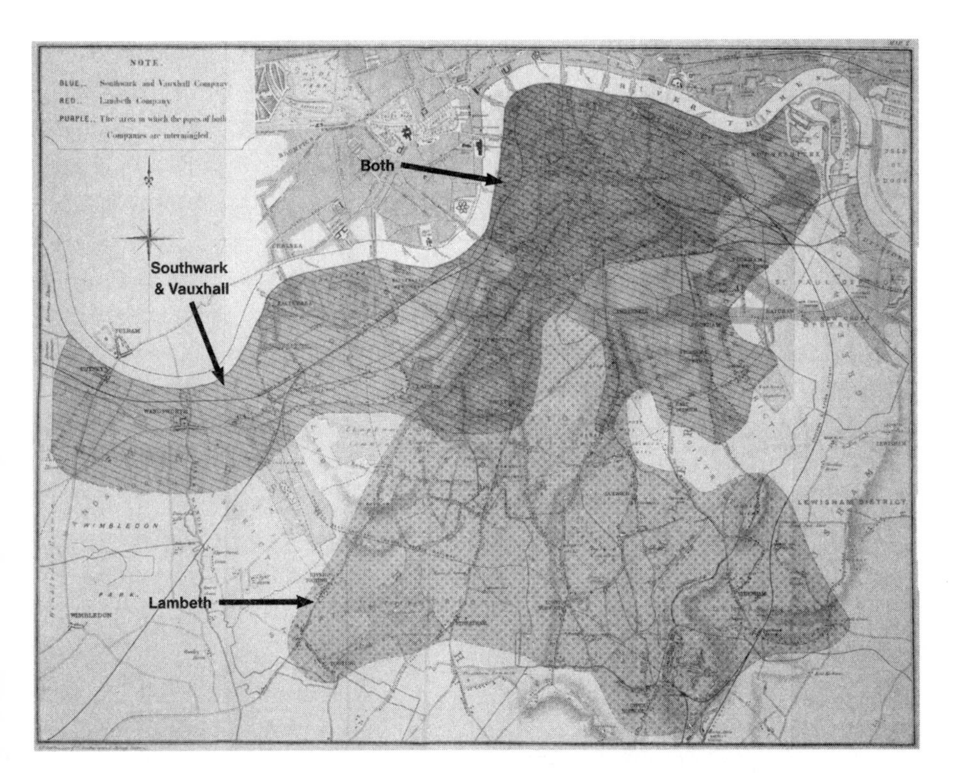

Figure 7.1 Areas of London supplied by two water companies, from the second edition of *On the Mode of Communication of Cholera* by John Snow. Water provided by the Lambeth Waterworks Company corresponded to much lower cholera rates than the water provided by the Southwark & Vauxhall Waterworks Company.

condition," as Snow described it, straight back to its customers. Lambeth, on the other hand, had recently moved their facilities upstream from the sewage discharge. Lo and behold, areas served by Lambeth had a much lower cholera incidence than the ones with water provided by the competitor. While this strongly suggested to Snow that cholera was spread through contaminated water, it still wasn't enough to convince the miasmatists. They pointed to other causes, arguing that residents of different districts were breathing different air, and that their living conditions could vary in other ways too. For Snow, this argument didn't, erm, hold water, and to prove it, he conducted possibly the first ever natural experiment.

To rigorously prove causation, it isn't enough to show that different outcomes (cholera death rates) coincide with different factors (water suppliers), as there might be other factors in play. What if residents of S&V-supplied

areas were poorer than the ones with water provided by the Lambeth company, and perhaps this could explain worse health outcomes? What if they were older on average, which would make them more prone to disease and less likely to survive illness?* Snow's observation proved only a correlation between water suppliers and cholera death rates; to prove causation, he needed more evidence.

He found a great opportunity in the area on the map where the two companies overlapped. In a part of South London, neighboring houses were supplied by different companies, and this was the only relevant difference between such houses: "[e]ach Company supplies both rich and poor, both large houses and small; there is no difference either in the condition or occupation of the persons receiving the water of the different Companies." In other words, people similar in most aspects were assigned to one or other of the two water companies "without their choice, and, in most cases, without their knowledge." If Snow managed to prove that his hypothesis held in this overlapping area, he'd have strong evidence that cholera spread through contaminated water. So, when cholera struck again in July 1854, he started knocking on some doors.

Armed with the addresses of cholera victims from the General Register Office, Snow went door to door, asking who supplied water in the house of each of the deceased—and he learned next to nothing. Often, the bills were paid by landlords who lived somewhere else, and even those who dealt with their bills themselves rarely remembered the name of the water company. (I, for one, wouldn't be able to tell you the name of mine.) To save the mission, whenever the residents couldn't provide him with the information, he'd ask for a water sample, in which he'd later measure the salt content. From this, he could identify the water supplier since he'd figured out that S&V water contained about four times more of this compound than Lambeth water.

The results were striking: by mid-August, thirty-eight out of the forty-four deaths in the area had occurred in houses supplied by S&V. With such promising data, Snow would have needed maybe just a few more weeks to finish up his grand experiment, as it's called today—but then, disaster struck.

*A similar argument partially explained why countries with relatively old populations, like Italy and the United States, observed so many deaths from COVID-19 in early 2020.

Detective Work

Snow had been pursuing his grand experiment for almost a year when the epidemic appeared in his neighborhood—Golden Square in central London's Soho. Being a scientist, he must have been tempted to continue with his experiment—he was so close to proving his theory!—but the local outbreak was more pressing. He suspected that the culprit was the Broad Street pump, so his priority became to act fast to prevent more deaths. Although his examination didn't find any impurities in the pump water, he couldn't think of other reasons for the outbreak. So, he dropped the experiment and started knocking on doors again, but this time, in a place he knew inside out.* After comparing the data from his on-ground investigation with the list of cholera deaths that occurred in the neighborhood in the week ending September 2 that year, a pattern emerged.

In his mind, Snow started seeing each death as a point on a map of the Broad Street area.† For almost all victims, the closest water source was the said street pump. Only ten deaths occurred in houses closer to other street pumps, but Snow was able to explain most of those cases too. Families of five of those victims told Snow that their loved ones would always bring home water from the Broad Street pump because, ultimately to their demise, they had preferred its taste. Another three victims were children who attended a school near the suspected pump and liked to take a sip of its water along the way.

What about the people who lived close to the Broad Street pump but didn't contract cholera? For example, out of the 535 residents of the workhouse on nearby Poland Street, only five died of cholera.‡ Snow promptly visited the institution and learned that they owned a private pump. Then there was a brewery, where none of the seventy-plus employees had died of cholera. The owner explained that they mostly drank malt liquor and, in the unlikely case they would prefer water, they could use their private deep well. In forty-eight hours, Snow managed to explain almost all such outliers

*Today's Broadwick Street is a popular shopping and dining destination. When my editor picked this neighborhood for our first lunch together, I was glad they were no longer experiencing water problems!

†The physical map wasn't drawn until late 1854 when the outbreak was over—and not by Snow himself but by a lithographer named Charles Cheffins.

‡In the United Kingdom, workhouses were institutions providing accommodation and (enforced) employment for the poor.

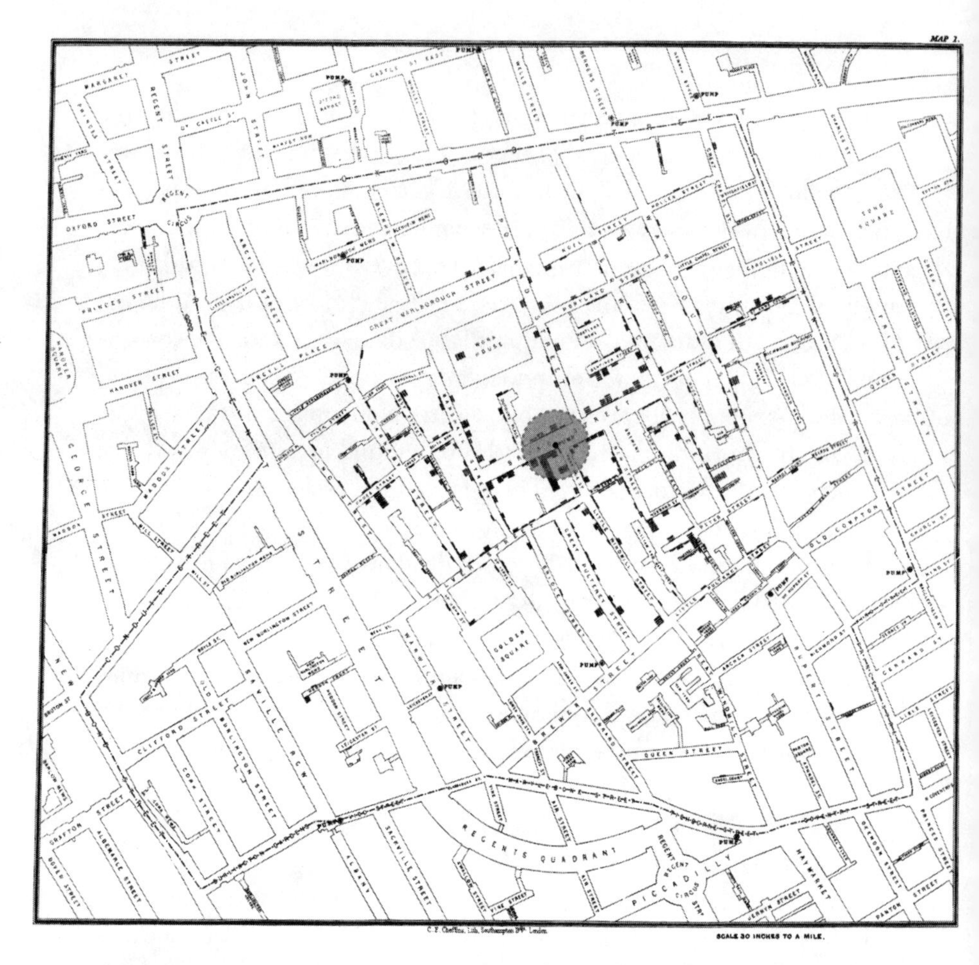

Figure 7.2 The famous map of cholera deaths (*black squares*) during the 1854 London outbreak, based on John Snow's data. The pump on Broad Street (*circled*) turned out to be the source of contaminated water.

by running back and forth between the houses and comparing the information he gathered with the data from the Registrar General Office. By then, Snow was certain that the Broad Street pump was to blame.

By assigning each house to the closest street pump, Snow divided the map into distinct regions. Today we call such partitions of a plane Voronoi diagrams, commemorating the Russian mathematician who studied them in detail, Georgy Feodosevich Voronoy. To create a Voronoi diagram, we define a set of locations on a plane—for example, water pumps on the map of Golden Square. Then, we look at the remaining points on the plane—such as homes of cholera victims—and for each one, we choose the closest (in terms of a chosen metric) location from our predefined set. This way, we divide the whole plane into regions, each containing exactly one of the relevant locations, and all points for which the location within the region is the closest one—in Snow's case, each of the regions contained one water pump and all the victims for whom that pump was the closest water source. The informal use of Voronoi diagrams helped Snow identify the outbreak's culprit. Modern applications of this clever technique range from preventing forest fires to understanding patterns on animal coats to assigning children to the closest school.

The Aftermath

On the evening of Thursday, September 7, Snow explained his case to the local authorities. They doubted his arguments, still convinced that it was the bad air that had caused cholera. Besides, the residents used to praise the high quality of the Broad Street well water. On the other hand, if Snow was somehow right, they could save many lives. After weighing the two options, the following morning the local authorities ordered the pump's handle to be removed. Soon, Soho was cholera-free.

After the outbreak, the pump got its handle back, and investigations into the Broad Street epidemic continued. A three-man committee appointed by the National Board of Health would spend months trying to understand what had happened. On paper, they were supposed to find the cause of the outbreak; in reality, they were trying to prove the miasma theory. But the person who truly got to the bottom of it—quite literally—didn't represent the government, but the church.

A young local curate, Henry Whitehead, had spent most of the outbreak with the dying. Disturbed by the tragedy, he too had set out to understand the epidemic. He wanted to debunk the prevailing theories explaining the outbreak, including Snow's argument that cholera was spread through contaminated water. But as he gathered data himself, he realized that Snow was right. In the end, Whitehead was the one to identify exactly where the contaminant of the Broad Street pump came from.

Considering cholera's few-day-long incubation period, Whitehead calculated that the first person would have gotten the symptoms around August 28. While studying the data, he encountered a note about a baby girl who had died on September 2, having battled with cholera for four days, longer than any adult. This made her the only candidate for patient zero, so Whitehead went straight to 40 Broad Street to talk to the baby's mother, who knew and trusted the curate. She recalled that while her child was sick, she'd thrown her soiled diapers into a cesspit by their house. The cesspit was promptly examined, and it was discovered that its contents were able to leak straight into the Broad Street well, which is how a little girl's diapers started one of the worst epidemics Soho had ever seen.

A Hero or a Storyteller?

Books, journals, and even medical textbooks have perpetuated the myth of John Snow the hero who single-handedly stopped the cholera epidemic. He has his society, based at the London School of Hygiene and Tropical Medicine,* which organizes annual Pumphandle Lectures and even reinstated a replica of the famous pump (without the handle) in the original location. A short walk from the pump, thirsty lecture audiences can have a pint in the historical John Snow pub. Almost two centuries after his research, Snow is still celebrated. But he has also had a lot of critics, both among his contemporaries and even today. We have no way of understanding the real impact of the removal of the pump handle, which prevented people from drinking water from the Broad Street well. Likely, the epidemic was already on the decline, and it would have ground to a halt in the area with or without Snow's intervention. The element of doubt means that we simply cannot support any claims that Snow saved hundreds of lives. Of course, this conclusion doesn't make Snow's study of the mechanisms of cholera any less valuable. Even if he didn't have any impact on the epidemic, his understanding that cholera spreads through water contributed to future improvements in sanitation.

Though typically presented as the lonely campaigner against miasma theory, John Snow wasn't the only person to suggest water as the mode of disease spread. His contemporary, a physician named William Budd, had a similar hypothesis based on his studies of typhoid fever. When in 1849 he took control of the water supplies of Bristol, in the southwest of England, he persuaded the authorities to invest money in a system of aqueducts that would bring drinking water from the uncontaminated rivers miles away

*In case you're interested, life membership costs £15 plus the price of shipping the special John Snow mug.

from the city. Why doesn't he have a society and a pub named after him? My best explanation is that we're drawn to a good story, and Snow had a knack for engaging his readers. In other words, it might be that we, the writers, are to blame.

Look more closely at Snow's story and you'll find that he is also falsely credited with founding the field of epidemiology, which is the study of disease patterns. Mapping fatalities as a way of tracking down the source of disease had been around at least since Valentine Seaman's 1795 map of the yellow fever outbreak in New York City. By Snow's time, medical mapping was so widespread that, in 1844, patients of the mental asylum in Glasgow killed time working on a map of a recent influenza epidemic. So, Snow wasn't the pioneer of presenting disease outbreaks visually; on the contrary, by the mid-nineteenth century, mapping spatial data had already gone mainstream.

Snow's research itself had many flaws, which most articles and books tend to pass over. His reasoning was along the lines of "the map shows many deaths around the Broad Street pump; therefore, the pump must be the culprit—case solved." He might, for example, have strengthened his argument by comparing the mortality around the Broad Street pump with mortalities in areas close to other pumps. While Snow didn't have the sophisticated statistical tools epidemiologists can use today, his analysis could—and should—have gone one step further. It seems that he rushed the publication of his findings, as evidenced by multiple versions of the same article released across a short interval of time, which left his critics unconvinced. It didn't help that Snow barely addressed the competing theories, presenting mostly arguments supporting his water hypothesis and ignoring any others. It's by finding evidence *against* alternative ideas rather than in favor of the researcher's theory that science pushes forward.

Why have I written at such length about John Snow despite all these criticisms, I hear you ask? Although his work wasn't flawless, and he didn't do it all alone, in my opinion, he still tipped the scales from miasma theory to our current understanding of waterborne diseases. If his greatest strength lies in storytelling, I feel that it speaks in Snow's favor. An engaging science communicator can contribute to scientific progress more than even the most gifted scientist who hides her or his research from the public. Science matters if we know about it and can act on it—and by "we," I mean "you and I," and not a handful of scientists in a lab. We should all raise a glass to the imperfect but fascinating John Snow and his map!

Beyond Cholera

Today, maps and mathematics are indispensable tools for fighting and preventing diseases, from local outbreaks to global pandemics. Scientists combine modern technologies from GPS to machine learning to gene sequencing with old-fashioned patient interviews to gather spatial information about the disease. In early 2020, many of us developed a daily habit of checking maps of COVID-19 cases, updated as the virus terrorized Wuhan, then other regions of China, Italy, and, finally, the whole world. Albeit unprecedented in scale, these weren't the first disease-mapping efforts in recent years.

Before taking potentially lifesaving decisions regarding vaccinations, quarantine, and social distancing, authorities need to understand likely scenarios of an epidemic's course under various circumstances. To predict how the infection might spread, scientists use mathematical models calibrated with the available data from the ongoing as well as past epidemics. Importantly, with today's powerful computers we can use these models to predict how the epidemic might evolve under different scenarios. For example, they can show what is likely to happen if no vaccines are administered, if half of the population is vaccinated, or if nearly everybody gets a vaccine. This information is crucial for authorities responsible for public health interventions.

The backbone of modern epidemiology is so-called compartmental models, which became popular in the early twentieth century. Some of the simplest are SIR models, which divide a community into three compartments describing their state of health: susceptible (S), infectious (sometimes called infected, I), and removed (sometimes called recovered, R). Individuals can transition between the states in two ways. First, a susceptible individual might become infectious if they come in contact with an infectious person. Second, an infectious person will recover and gain immunity from the disease, or die; in both cases, this individual will be removed from the infectious compartment. In more complex models, the assumption that a recovered individual gains a lifelong immunity to the disease—which doesn't hold for all diseases, as every person who has gone through COVID-19 more than once has painfully learned—can be altered.

The model is written as three equations, each describing the number of individuals in one of the compartments at a given time after the beginning of the epidemic. The equations include parameters expressing the transmission rate and the recovery rate, which impact the probabilities of individuals transitioning between the compartments. For example, the longer an individual is infectious, the more likely she is to recover. This simplistic SIR model has many variations, which might include additional states or socioeconomic data about the population, for example the age structure. It can also account for public health interventions such as quarantine, and this account is useful in generating possible scenarios of the disease's spread. All compartmental models, however, treat the population as a single community closed off from the rest of the world and ignore any geographical characteristics.

In our interconnected world, it's hard to contain a disease, which is where models with a spatial dimension come in. Such models often describe the epidemic both on micro and macro levels. On a micro level, they use a variation of the SIR model to describe the spread of a disease in a small community. The macro level describes the potential spread of the disease between communities through a transportation network.

One of the most widely used examples is the Global Epidemic and Mobility (GLEAM) model, successfully used to understand, for example, the 2009 H1N1 pandemic. The model uses real-world population and mobility data. The available population dataset divides the world into small regions of 15 by 15 arc minutes,* which along the equator corresponds to about 17 by 17 miles (the measurements vary depending on the latitude), and provides the number of residents of each region. To understand how people move between these small regions and coincidentally spread diseases, researchers had to connect this dataset with information about passenger flows between major transportation hubs. For that, they assigned each region to its closest major transportation hub, which might be a major city or an airport. This created a Voronoi-like diagram of over 3,200 communities concentrated around transportation hubs, similar to the one in Figure 7.3.

The disease dynamics within each community are described with an SIR model. In addition, GLEAM contains equations governing the mixing of these subpopulations, either via everyday commute or long-distance traveling. This means that the number of infectious individuals in a given subpopulation—the Pittsburgh area, let's say—depends on the numbers of infectious individuals in all other hubs. But not all hubs impact Pittsburgh's

* An arc minute is equal to one sixtieth of a degree of latitude or longitude.

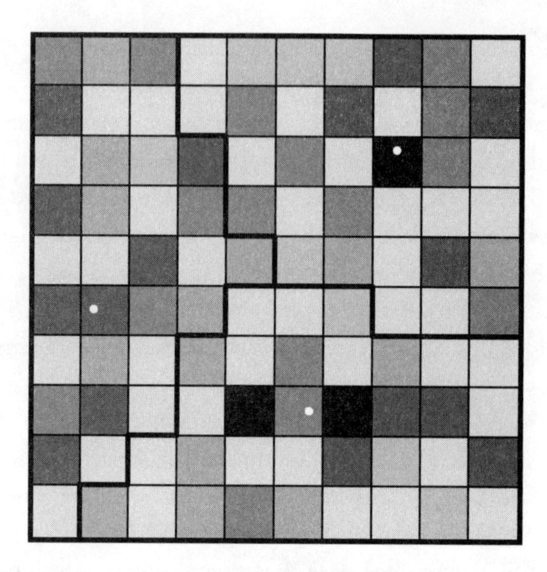

Figure 7.3 Each cell of this Voronoi diagram contains one major transportation hub and all small regions for which it is the closest hub. Darker regions have larger populations.

subpopulation in the same way: the dependence is stronger for strongly connected hubs. For example, infectious residents of nearby Cleveland are more likely to travel and infect someone in Pittsburgh than residents of more remote Nashville. In practice, this means that the number of susceptible people in our Voronoi cell on day 2 is equal to the number of susceptible people on day 1 minus the number of people who got infected between day 1 and 2, plus the number of susceptible people from other Voronoi cells who arrived to our cell between day 1 and 2, minus the number of susceptible people who left our cell between day 1 and 2. We construct equations describing the numbers of infectious and removed individuals similarly. In addition, not all individuals in one hub have the same probability of traveling or commuting. So, the model accounts for the fact that sick people are less likely to travel than healthy ones.

Without mathematical models and maps, we'd be all but helpless against infectious diseases. Maps and math help us identify likely sources of outbreaks and predict the likely evolution of the epidemic, but they're just tools, not magic wands. During a long and fruitful conversation, medical geographer and ethicist Tom Koch tells me that he uses maps to explain diseases for a living but is the first to admit that no single map can tell us the whole story. What a good map can do, however, is suggest a hypothesis so that "we go and look at more data, which may be able to complete the story." It's all about asking the right questions, finding the right data, and using the right tools.

These observations apply not only to public health but also to whenever we want to make a spatial argument. Crime investigation, for example, has a lot in common with tracking diseases: there's the offender, victims, evidence, and detective work. Epidemiologists and police investigators use similar tools and heavily rely on maps and mathematics.

The Police Officer Who Loves Math

One morning at 6 a.m. I get on a video call with Kim Rossmo, director of the Center for Geospatial Intelligence and Investigation in the School of Criminal Justice at Texas State University. Upon seeing my sleepy expression, he assures me that we could have agreed to talk later, even though it's 11 p.m. for him. "I'll be up later," he claims, "too many night shifts." I guess he's referencing his decades in the police force.

For Rossmo, the road to investigative fame wasn't straight. As a teenager, he was interested in chemistry but gave up on a career as a chemist after cleaning too many test tubes. Math required just pen and paper, which was more appealing, so it became his university major. Rossmo's excitement quickly waned as he realized that the lectures he was required to attend were mostly about proving theorems and not about applying math to something useful. After two years, he took a year off from his studies to think about what he wanted to do. At that time, he was dating a psychology graduate student who would practice her craft by administering vocational aptitude tests to him. Rossmo would always come off high on "adventure" and "excitement" as well as "investigations," which led him to criminology.

When he came back to finish his undergraduate degree, Rossmo switched from math to sociology, as it was the only degree to offer criminology modules. On top of his studies, he worked as a detective for a local security company. Rossmo loved both his new major and his night investigations, so, after graduation, he applied for jobs at police departments all over the world, carefully choosing cities that offered graduate programs in criminology, just in case he wanted to enroll in one in the future. He was almost accepted by the Vancouver Police Department in his home country, Canada, but he didn't satisfy one requirement: he didn't know how to swim. So, he learned, and kept his job there for almost twenty-three years.

After completing three years of training, Rossmo took postgraduate degrees in criminology at Vancouver's Simon Fraser University: first an MA, and then a PhD, which he worked on following a long day (or night) on duty. In 1995, he became the first police officer in Canada to be called a Doctor of Criminology. Rossmo's doctoral research has given us a tool to capture serial criminals—and save lives.

The Geography of Crimes

During his graduate studies at Simon Fraser University, Kim Rossmo was introduced to the work of environmental criminologists Patricia and Paul Brantingham. Environmental criminology focuses on the "where" instead of the "why" of the crime; that is, it studies maps and locations rather than the psychological and socioeconomic characteristics of the criminal. The main idea of the field is the obvious but crucial condition for a crime to occur: the criminal and the victim must find themselves in the same place at the same time. In most homicides, this condition is naturally fulfilled, as the victim usually knows the offender. Serial killers, however, tend to pick their victims at random, based only on the location and additional factors such as the victim's age or gender, which makes serial offenders more difficult to identify.

The Brantinghams have devoted their careers to studying the spatial aspect of crimes. While each crime is unique, they all follow some general patterns. Criminals hunt for victims in places they know best, for example, close to home or the workplace. At the same time, they avoid attacking too close to home, creating a "buffer zone." Together, these two characteristics create a Goldilocks zone where the criminal feels comfortable. We all have activity spaces—places we live, work, and visit regularly for whatever reason—and so do criminals. The Brantinghams' model helps predict the approximate locations of future crimes given the activity area of the offender.

When they came up with their theory, the Brantinghams had in mind crime prevention—by understanding what constitutes a risky area, we can increase policing, improve street layouts and lighting, or introduce other preventative measures. Rossmo wondered if one could flip this approach around to learn valuable information about serial criminals. Could the locations of multiple crimes, together with the assumptions of environmental criminology, tell us something about the offender? Rossmo often brings up the analogy of a rotating lawn sprinkler: we can't predict where exactly a droplet will land, but based on the landing locations of many droplets, we can figure out the sprinkler's position. Similarly, we could use the crime locations to approximately identify where the offender lives or works. And that's how geographic profiling came to be.

Of course, identifying a serial offender is more difficult than finding the source of water droplets on a lawn. First, while we can reasonably assume that all the water droplets have come from the same sprinkler, we can't be sure which crimes have been committed by the same person. This requires some Sherlock Holmes-y thinking: Were many similar crimes committed in a short period in a neighborhood close by? Did the offender leave some "signature"? Or maybe we're fortunate enough to have found matching DNA in multiple locations? With more complex cases, the investigators analyze multiple scenarios, adding or removing a dubious crime from the dataset—especially when the crimes have been committed over a long period or in a large area.

After the offenses have been assigned to a single culprit, the serial crime investigator often still doesn't have many data points to use. While thousands of water droplets lead to the sprinkler's location, one offender—fortunately!—rarely commits more than a couple of crimes. Investigators have some clever ways to increase the number of locations to look at, as it's not only the crime locations that are relevant but all locations where the offender has visited. In the first geographic profiling case Rossmo was involved in, for example, a serial rapist used a stolen car to get to his victims. Having found the owner's credit card in the vehicle, the criminal used it liberally, letting the investigators add purchase locations to the dataset.

If we are to believe TV shows, we can expect geographic profiling to put a single "X" on the map, which would be followed by a spectacular police raid on the criminal's residence. But Rossmo didn't create his method to solve crimes, that is, to identify the offender. Instead, his goal is to narrow down the geographical search area, which, together with other pieces of evidence, can lead the investigators to the suspect. Geographic profiling is one of many tools available to the police—a tool at once powerful and limited.

The Train of Thought

Geographic profiling assumes that each time a crime is committed, the offender leaves the home base searching for a victim and returns there afterward. If this key condition isn't satisfied, Rossmo's method won't work. If the investigators suspect that the criminal changes their home base between crimes—for example, if they travel through the country, committing crimes along the way—geographic profiling might not be the best tool to use. Neither will the method give satisfactory results in cases where the offender doesn't search for victims geographically; for example, if they instead choose victims over the internet.

Given a set of locations associated with the offender, geographic profiling determines their most likely base. "Most likely" is the key: the analysis won't return the offender's address but will identify areas where the investigators should focus their search. This is why it took a person with mathematical knowledge to develop the method.

The goal of geographic profiling is to divide a map of the investigation area into small squares and assign each of them a probability that it contains the criminal's base. To compute these probabilities, one uses Rossmo's formula, as it's known today. The formula consists of two main parts: one responsible for the areas outside the buffer zone and one inside it. The first term is zero inside; outside the buffer zone, it expresses that the probability decreases as the distance from the crime location increases. The second term is zero outside the buffer zone and assigns increasing probabilities as the distance from the crime location increases within the buffer zone. When summed, these two terms describe the Brantinghams' model in mathematical language.

The sum of the two terms is computed for all crime locations suspected to have been committed by the same offender. Then, the resulting sums from all crime locations are aggregated and scaled appropriately to ensure that each final probability value falls between zero and one. The resulting geographic profile is a map of the investigation area with each point colored according to the probability of containing the offender's base,* from unlikely areas in blue and purple to most likely locations in orange and red. The result resembles a topographic map containing a volcano with a caldera: the peak in red, corresponding to the area of highest probability, and the middle in yellow, representing the buffer zone. It's an interesting coincidence that Rossmo came up with this idea passing by the volcano Mount Fuji on a bullet train in Japan.

* More precisely, a rectangular grid is overlaid on the map, and each cell is assigned a probability of containing the offender's base.

Rossmo frantically scribbled down his ideas on Japan Railway napkins, which raised some eyebrows among fellow passengers. He came back to Vancouver with a bunch of napkins and a sketch of an algorithm that now needed implementation.* As he was not a programming expert, he needed nine months and a little help from an engineer friend to write the first geographic profiling computer code. A trial run on an arson case gave accurate results, which encouraged Rossmo to continue the research that became his doctoral thesis. His work received a lot of media attention and he got—as he admits himself—"a little addicted" to the positive response for doing something useful. So, he's been researching, applying, and teaching geographic profiling ever since.

The Chair Burglar

It's the early 2000s, and a burglar has been breaking into houses in the city of Irvine in Orange County, California, for years without being caught. The police believe that all the relatively wealthy houses were broken into by a single professional burglar who approaches the job systematically. The offender only breaks into single-family homes, mostly those neighboring a park, meaning no trespassing is needed to plan the heist. Having stolen some cash and jewelry, the burglar always exits at the back of the house, often using a chair to climb over the fence. It's because of the latter habit the mysterious criminal became infamous as the "chair burglar."

Even the best burglars make mistakes, and the police find DNA traces in five of the burglary locations, three of which belong to the same male. Catching the offender red-handed would be all but impossible, given how quickly he's able to get into and out of a property, so identifying the suspect while he is planning the next burglary seems like a better bet. The area the burglar has covered so far, however, is way too large for constant policing: it's about three-quarters the size of Manhattan. The police want to catch the chair burglar while he's choosing his next target, but it isn't possible to patrol the whole area of about seventeen square miles. So, the Irvine Police Department's in-house geographic profiler Lorie Velarde gets straight to work.

* Remember, an algorithm is a set of instructions, usually for a computer.

The multitude of burglaries that has already been committed created a rich dataset from which to generate a geographic profile of the chair burglar. Velarde manages to shrink down the search area about seventy times, narrowing it down to around a quarter square mile most likely to contain the criminal's base. Usually, the base is the offender's home or workplace, but Velarde figures that a professional burglar isn't likely to live in the wealthy neighborhood of Irvine, so he probably commutes to work from outside the community. (Organized stealing is a job, too, apparently.) Under that assumption, the peak of the geographic profile corresponds to the starting point of the burglar's house hunt when he arrives in Irvine. This observation allows the police to focus on cars with nonlocal license plates, whose owners are later identified through the California Department of Motor Vehicles.

On the first evening, officers observe a nonlocal car in the peak geoprofile area, which turns out to be a rental car from Los Angeles County. The police don't stop the car, but they learn from the rental company that Raymond Lopez, who happens to be driving the car that evening, is among their best customers. He has been renting from the company for two decades, each time insisting on a different car. Not suspicious at all . . . When Lopez returns another car to the rental company, the police find his DNA on the steering wheel. It matches the sample previously collected at burglary sites, which warrants his arrest. He's found guilty of 139 burglaries in two years alone, with a property loss of over $2.5 million, as well as possession of a handgun. One can't say he wasn't good at his job!

Geographic profiling has been applied not only to investigate crimes but also to counter terrorism, locate seventeenth- and eighteenth-century pirate ship bases in the Caribbean, determine earthquake epicenters from digital reports, and even research hunting patterns of white sharks. Today, Kim Rossmo is researching the possibility of applying his method to search for missing persons in the wilderness based on the signals from GPS trackers that most of us carry in our pockets.

Various approaches to the search for lost people and objects have been evolving for decades. One of them—conceptually similar to geographic profiling but instead based on a centuries-old theorem—has proven particularly successful. While Bayesian search, as it is called, sounds rather obscure and mathematical, I suspect not only that have you heard about this method in the news but that you're using it yourself on an everyday basis.

When Air France Flight 447 from Rio de Janeiro to Paris sent its last GPS position just after 2 a.m. UTC on June 1, 2009, it became the last known position before it tragically disappeared into the ocean, taking with it 228 passengers and crew members. For the next two years, the whole world followed attempts to recover the wreckage, which would help to explain what had happened and give the mourning families much-needed closure. And the whole world wondered how difficult it could be to spot a huge Airbus. Very difficult, it turns out, when we consider the vastness of the Atlantic Ocean—so maps and math were needed again.

The Most Useful Theorem

If I had to choose what, for me, is the most useful piece of math, it would be Bayes's theorem. Formulated in eighteenth-century England by the Presbyterian minister Thomas Bayes, it tells us how to update our beliefs as we gather new information, which makes it particularly useful when searching for lost objects—as big as plane wreckage in the ocean and as small as misplaced headphones in my one-bedroom apartment.

To find my headphones, I could frantically turn my apartment upside down; alternatively, I could apply a methodical, Bayesian search. Are they in the living room? The bedroom? The bathroom? The kitchen? Before I start searching, I need to formulate the initial probabilities of the headphones being in each of those rooms. One way would be to look at the locations of headphones I, or people like me, have misplaced in the past. This data-based approach would give me objective probabilities—but to have such a dataset I'd need to regularly misplace headphones and meticulously track their ultimate location. As a mathematician, I've done a lot of weird things, but this is *not* one of them.

I can, instead, analyze my habits and experiences to come up with subjective probabilities.* For example, I tend to listen to podcasts while I cook, but not when I hang out in the living room, which makes the kitchen a more likely location for my headphones. I'd say that my subjective probabilities look like a chance of 60 percent for the kitchen, 20 percent for the living room, 10 percent for the bedroom, and 10 percent for the bathroom.

*Some statisticians don't like to use subjective probabilities—in my humble opinion, for no reason. As long as the initial probabilities are somewhat reasonable—for example, they stem from existing scientific knowledge or our past experiences—the method gives powerful results.

Because I'm sure that the headphones are *somewhere* in the flat, these initial probabilities add up to 100 and together form what is called the prior.

Now that I have my prior, I can start looking for the headphones in the room where my chance of success is the highest, that is, the kitchen. Alas, they are not here! To keep things simple, let's assume that I have excellent searching abilities—in other words, the failure to find the headphones in the kitchen means that they must be somewhere else, not that I have failed to spot them. So, the probability for the kitchen falls to 0 percent, and the 60 percent gets equally split between the other three rooms, giving 40 percent for the living room, 30 percent for the bedroom, and 30 percent for the bathroom (notice that the probabilities still add up to 100 percent). These new probabilities form what is called a posterior. Now, I'll repeat the process with the room that currently offers the highest probability of me finding the headphones, and each time I search I will update the posterior until I find them. If, by the end of the search, I come away empty-handed, I'll have to re-evaluate my assumptions—either I've lost the headphones outside of the apartment (which would be unfortunate) or my ability to find lost objects isn't perfect. That is, I might have missed them while searching the room they were in.

Looking for plane wreckage is much more complicated, but the principle remains the same. Crashes happen extremely rarely, which is lucky for humanity, but unlucky for the investigators who can't work with objective probabilities simply inferred from an extensive dataset of historical accidents. Instead, they must rely on subjective, but still evidence-based, probabilities.

Usually, a large search area will be divided into a rectangular grid, with all cells assigned probabilities of containing the wreckage adding up to 100 percent. While I might be sure to find the headphones if they're in a room I'm searching, even a huge plane can be overlooked in a deep, dark ocean. So, for each grid cell the researchers estimate how likely they are to find the sunken object if it's there and, using this information, they adjust the probabilities. These values depend on many factors, but the most important is the water depth—the deeper it is, the less likely the chance of success. Then, the search starts in the areas of the highest probability, moving on to the areas of lower probabilities, until the wreckage is found, or the probabilities become too low to justify the search costs. After each unsuccessful search phase, the probabilities are updated with the new information. The information might simply be that the wreckage hasn't been found in the grid cell, but it also might include conclusions from, for example, finding the bodies of victims or pieces of the plane.

Bayesian search may seem like an application of common sense rather than a groundbreaking idea, but describing our intuition in mathematical language can be surprisingly powerful. If we rely only on common sense, the search becomes disorganized and has a lower chance of success. Importantly, Bayes's theorem allows us to incorporate all pieces of information, even the conflicting ones, which prevents human investigators from clinging to their initial idea.

Aware of the power of Bayesian search, French officials asked the American consultancy Metron for help. For decades, the company's researchers had been applying mathematics to methodically search for objects and people, from the gold-loaded shipwreck of SS *Central America,* which sank in 1857, to the submarine USS *Scorpion,* which disappeared in 1968. Metron's search optimization technique has also saved dozens of lives after the team developed the Search and Rescue Optimal Planning System (SAROPS), used by the US Coast Guard to search for sunken boats and people lost at sea.

As soon as French officials decided to give Bayesian search a try, Metron's expert Colleen Sterling flew to France to help solve Air France's case. Curious to hear about her experiences, I too did some detective work to contact her—and here we are, chatting about plane crashes shortly before I'm due to take a transatlantic flight.

How to Find a Missing Aircraft

A pilot and aircraft owner herself, Sterling knows all too well what it's like to be up there in the sky, vulnerable to the caprices of the weather, flawed equipment, and one's imperfect decision-making. In her small aircraft, Sterling has spent hours flying up and down the California coast, where she lives and works, and other locations all over the United States. While this time the stakes were higher than usual, the Air France wreckage wasn't the first missing aircraft she'd tried to find.

In September 2007—almost two years before Air France's crash—a well-known adventurer, Steve Fossett, had failed to return from his short Labor Day flight in the Lake Tahoe area. Sterling watched the search with interest. "As airplane owners, we're always interested in why people crash so that we don't do what they did," she explains. As the rescue team came back empty-handed again and again, Sterling couldn't help but observe how chaotic this search was in comparison with Metron-supported, organized rescue missions. When she suggested to Lawrence Stone, Metron's CEO at the time, that they should help, he warned her not to get involved in overland search and rescue because "it's really messed up."*

She persisted, however, and finally Stone offered her Metron funding to work on the case in her spare time, under the condition that she obtain some data and get someone in charge of the current operation to listen to her. By then, the Sierra Nevada Mountains, where Fossett was likely to have disappeared, were covered with snow, so the search efforts had ceased—at least the official search. From her many connections in the local aviation community, Sterling learned that a team of Fossett's friends would keep working over the winter. She discovered that they were aware of Bayesian search but had no idea how to apply it.

Over the next few months, Sterling worked with Fossett's friends, preparing probability maps, doing computer simulations, and analyzing radar tracks. They classified parts of the terrain according to search difficulty, from relatively easy open areas to all but inscrutable mountainous terrain. They looked at all the search efforts to date and, based on their analysis, proposed areas that should be searched again. At some point, Steve's wife, Peggy Fossett, funded a week of use of a "fancy helicopter," as Sterling describes it. Alas, even the expensive equipment didn't help. In the summer, Sterling joined Fossett's fellow explorers who were camping in the mountains of Nevada for two weeks, searching the canyons and the mountainsides, and talking to people who had searched the lakes—all in vain. Finally, about a year after Fossett's disappearance, a hiker found his ID card and a bunch of $100 bills close to Mammoth Lakes in the Sierra Nevada. The next day, the wreckage of his plane was discovered in that area.

Finding the aircraft might have concluded the story for most, but not for Sterling. She was disturbed by the fact that the wreckage turned up in an area barely searched by the rescue team, so she wanted to understand where they had gone wrong. She summarized her findings and suggestions for what could improve future searches and started going from one conference to another with her presentation. To Sterling's surprise, agencies whose

*According to Sterling, he used a less polite word, but it wouldn't pass my editor's scrutiny.

job is to deal with aviation emergencies didn't want to listen to her, even though she made sure to focus on improving the future rather than criticizing the past. It seemed that everyone was afraid to lose their funding should they admit to any mistakes, even if correcting these mistakes could be a matter of life and death.

One of the main culprits of failed search missions is the lack of digital records of searched areas. In Fossett's case, the official, government-funded search team hung up a big printed map on which they'd shaded in areas that—in their subjective opinion—they'd searched thoroughly. Instead of the Bayesian approach of combining rigorous estimates of the search quality into the overall probability of finding the target, however, these were guesstimates from the pilots who had looked at the area. And pilots—like most humans—tend to overestimate their perceptivity. It became Sterling's goal to persuade search teams to assess the search's quality in two ways: by collecting GPS tracks and more realistically calculating the search effectiveness after each search phase.* These two pieces of information, combined with the Bayesian theory, would help identify the next search area. Otherwise, one gets "distracted by bright new shiny things," Sterling says, like "somebody said they saw the airplane—let's run over to Nevada and spend a lot of time in Nevada." This is what happened in Fossett's case: the search team started following a radar track of his plane, but the next day they learned about a visual sighting and simply ditched the trace. We tend to give more weight to visual sightings, even though people are "really horrible at recognizing airplanes from the ground." Without a rigorous search plan, they never got back to the original radar track.

Metron had an opportunity to apply the insights gained from Fossett's failed search when the French government asked for help with Air France 447. This case was different—not only because of the scale but also because the French authorities understood the value of Bayesian search and had the money to pursue Metron's recommendations. Indeed, the two-year-long search mission cost tens of millions of euros and took efforts from multiple countries and organizations, but it ultimately succeeded soon after Metron brought math to the table.

*Search effectiveness is affected by factors such as the searchability of the terrain, the sun angle (Sterling asserts that "looking at the terrain into the sun is really tough!"), the speed and altitude of the search aircraft, and the number of observers on board.

Second Time's a Charm

The French Bureau d'Enquêtes et d'Analyses (BEA) organized an international search mission immediately after receiving the information about the crash. Based on the last known position of the plane and the next expected signal that never arrived, investigators determined the search area. The first days focused on exploring the ocean's surface, which uncovered dozens of floating bodies as well as pieces of the aircraft. But to find the black boxes, which would contain the crucial information about the crash causes, an underwater search was needed. Alas, despite multiple attempts, including a thorough submarine search, by late May 2010—almost a year after the disaster—no black boxes had been found. In July 2010, BEA asked Metron scientists for help analyzing the unsuccessful search efforts and creating an updated probability map to suggest the next steps.*

I learn from Colleen Sterling that the case came with a lot of restrictions attached. Metron was explicitly forbidden from assuming any problems with the aircraft or fault of the pilot—in other words, from forming a prior that could lead to lawsuits against Air France (lawsuits followed anyway). Instead, Metron included the possibility of human error implicitly, using the data collected from other crashes caused by pilots losing control of the aircraft.

They started by gathering all the available information about the accident's possible location. This allowed them to generate possible scenarios, quantify how likely each of them was, and generate a prior. The prior accounted for two main factors: the flight dynamics and the so-called reverse drift, which describes motion in water. The first factor helped estimate how far the plane could have flown between its last recorded position and the expected next signal, given the flight altitude and the data from past commercial aircraft accidents. The second factor involved analyzing the data of currents and winds in the area, which allowed the scientists to reverse-engineer the movement of bodies found in the ocean. In other words, oceanographers looked at each recovered body and asked where its journey was likely to have started. Because of the complicated system of

*Metron was first consulted a few weeks after the crash, in mid-July 2009, when they created an initial search map. I learned from Sterling, however, that Metron's initial search assessment was mostly ignored. Also, the power of Bayesian search lies in updating the probabilities with new information, which Metron wasn't asked to do until July 2010. So, the initial consultation didn't result in much success.

currents in this part of the Atlantic, however, these estimates carried a lot of uncertainty. Still, the beauty of Bayesian search is that they didn't have to discard this information but could include it in the prior with a relatively low weight. Metron also assessed the likelihood of a failed search; this step was important because hard-to-search areas likely to contain the wreckage could be worth revisiting.

With the prior ready, Metron scientists analyzed the phases of BEA's unsuccessful search, each time updating the probabilities and creating a new posterior. They started to question some of the assumptions that had guided BEA's yearlong efforts. In particular, they weren't convinced that the black boxes' so-called pingers were still functional at the time of the search. These devices exist to aid underwater search after a potential accident and are designed to survive the impact of a crash and activate in contact with water. According to the manufacturer, pingers should send signals for about forty days after the crash, but during BEA's underwater search, no such signals were detected. While the lack of signal led BEA search teams to assume that the wreckage wasn't there and to move on, Metron wondered if these pingers had been working in the first place. Although the data from past crashes suggested that their combined failure would have been unlikely,* Metron decided to create an alternative posterior, which assumed that the pingers had been destroyed in the crash. In that case, the lack of signal wouldn't necessarily mean there was no wreckage in the area, so the search results would need to be reinterpreted. It turned out that they were right: the pingers had not been working, making it all but impossible for BEA to locate the plane in the initial pinger search. Metron's unorthodox thinking led to success: in early April 2011, a few days after the search was resumed, a large part of the wreckage was discovered in a high probability area indicated by the new posterior.

The recovered flight data recorder and cockpit voice recorder were invaluable sources for determining the causes of the crash. The black boxes revealed what had happened in the last minutes of the tragic Flight 447. As a heavy storm hit the plane, ice crystals formed on its airspeed sensors, which disabled the autopilot. Pierre-Cédric Bonin, a young pilot, took control of the plane, which he wasn't used to flying at all. Automated systems

*The failure of one pinger is unlikely, and the two pingers work independently from one another. In such case, we multiply together these two already small probabilities, which gives an even tinier chance that both pingers fail.

are extremely reliable—and while that's a good thing, it also means that pilots don't have a chance to gain real experience steering the plane in unusual conditions. When the plane encountered a small amount of turbulence, it tilted gently. Desperate to level the wings, Bonin moved the control stick in the opposite direction so hard that the plane started rocking. By itself, this shouldn't have caused much trouble besides some spilt drinks in the cabin, but the panicked pilot also moved the control stick in the vertical direction. Confused by faulty readings from altitude sensors, the pilot pointed the aircraft's nose upward, steering the plane for a steep climb. The situation started to get a little dangerous, but by no means out of control—as the alarms in the cockpit sounded, all Bonin had to do was point the nose downward to level the plane. Instead, he kept moving the plane upward, which finally caused the plane to stall and then plunge into the water at a breathtaking speed. Ironically, reliable automated systems cost the lives of 228 people.

Thanks to the efforts of dozens of investigators equipped with Bayes's theorem, a lot was learned from this tragedy. Nothing will console the mourning families, but recovering the black boxes might at least have prevented similar tragedies and saved the lives of future passengers. Air France's crash led to improved pilot training; today, pilots get more experience controling the plane by hand.

For all the math and logical application of data, the jobs of Sterling and her colleagues are emotionally taxing. "I deal with it," she tells me, "by thinking that by working on these things, we can try to prevent these accidents, by understanding what happened. And you give closure to the families by finding their loved ones."

The Perfect Couple

When searching for people and things—from patient zero to criminals to sunken planes—it's tempting to frantically follow every lead that comes up. Following our instincts, however, won't get us far. As we've seen, it's the organized, mathematical search that gives us the best chance of success.

The use of maps in various searches isn't surprising; we've all read adventure books featuring treasure maps and, as children, we may have spent many a blissful summer afternoon participating in orienteering games. But maps alone won't tell us where to search first, which areas to visit more than once, and when we should change the strategy—for that, we need mathematics.

As we better understand the power of cartography and math working in tandem, we see that on the one hand, by visually representing statistical data, maps turn a bunch of impenetrable numbers into a clear picture that can give us new insights. And that on the other, when we apply the theory of probability to the map, we can optimize our search. Maps and math together save lives.

8

DEEP

How to Map the Invisible

Like many map nerds, I've spent countless delightful hours discovering the world through Google Earth. It feels reassuring to see that after centuries of cartographic developments, we finally have a complete map of both the continents and the oceans—or do we? As it turns out, we know less about the bottom of our oceans than about the surface of Mars. Google Earth's images don't represent direct measurements of the ocean depths, they represent satellite data—and these data are much less precise than we'd wish. For as much as 80 percent of the ocean floor, this is the best information we have, making the vast majority of the Earth's surface a true terra incognita.

To gather information about the Earth's surface, a satellite—more precisely, a satellite radar altimeter—sends a radio signal toward the Earth and measures how long it takes for it to return after it gets reflected from the Earth's surface. Since we know how fast radio waves travel in the air, we can compute how far the surface is from the satellite.* This method gives us precise estimates of altitudes on land as well as the sea levels, but since radio waves don't pass through water it says nothing about the shape of the ocean floor. That is, until we realize that large underwater mountains

* If radio waves travel in the air with speed v and the signal takes the time t to return to the satellite, then the satellite is at the altitude $(tv)/2$ above the Earth's surface.

increase the Earth's gravitational force, creating sea-level "bumps," while the valleys decrease it, creating sea-level "dips." So, satellite-based measurements of differences in sea level let us map large landforms on the ocean floor. These imprecise data make up most of what we see on ocean parts of Google Earth. Because radar signals quickly die off in water, mapping the bottom of the ocean is much harder than mapping dry land, even on other planets. While we've mapped most of the surfaces of the Moon, Venus, and Mars with a resolution of at least 100 meters, most of the seabed's maps have a resolution of only a few kilometers. This means that Google Earth and similar maps show only features with at least one dimension exceeding a few kilometers, omitting anything smaller.

Precise maps of the ocean floor are crucial for reasons beyond the safety of tens of thousands of ships large and small crossing the oceans every day. To build anything in the ocean—for example, wind turbines—we need to know what lies on the bottom. To understand the weather and our changing climate, we need to map underwater landforms that impact ocean currents. To protect people and habitats from natural hazards like tsunamis, we need a detailed knowledge of the seafloor. In short, to ensure the safety and prosperity of both the oceans and people, we need maps that show us more than only the largest landforms on the seafloor.

In 2017, two nonprofits—the Nippon Foundation and the General Bathymetric Chart of the Oceans (GEBCO)—launched the Seabed 2030 initiative with an ambitious goal of mapping the entire ocean floor by 2030. This goal aligns with the UN's Sustainable Development Goal 14, which is to "conserve and sustainably use the oceans, seas and marine resources." The project leaders encourage international collaboration between research institutions, governments, and even regular citizens to create a publicly available, complete, and detailed map of the ocean floor. Some of the data already exist, hidden in inaccessible databases, so the Seabed 2030 project aims to collect and publish all such information. Maps of the remaining parts of the world's oceans need to be created from scratch, which involves expensive and time-consuming data collection. Although such a project can't avoid sending out research ships with dozens of scientists on board, the technology has become much more efficient since we first began measuring the depth of the ocean floor, and the task, for the first time, feels possible.

What the Depth Sounds Like

Satellite-based mapping gives a rough estimate of the ocean floor, but to obtain more detail, scientists need to get their hands dirty or, rather, wet. For at least three millennia, depth measurements—called *soundings**—were done with a lead line, that is, a thin rope with a weight attached to one end. The navigator would lower the line from the ship until it hit the bottom, pull it back up, and measure the length of the rope that had been in the water to get the depth.† In combination with other measurements that established the ship's position, this whole process would give one point on the map of the seafloor. Sounding with a lead line was tiresome, time-consuming, and imprecise, but it was the best we could do from antiquity until the early twentieth century.

The 1912 *Titanic* disaster, followed by the extensive use of submarines during the First World War, inspired rapid development of methods for detecting underwater objects. Only then did we figure out how to use sound to "see" underwater, employing a technique that animals such as dolphins and bats have been using since time immemorial. Echo-locating animals send out sound signals, which, when reflected from an obstacle, return to the sender, informing them of what lies ahead. It's no different to calling out a "hello" in a large, empty room and waiting for our voice to echo back with a slight delay. Similarly, single beam sonar (short for sound navigation and ranging), patented in 1913, sends a sound wave toward the ocean floor and receives back a delayed signal. It works on the same principle as satellite radars, although it sends sound waves instead of radio waves. The depth of the ocean is equal to half of the time the signal's round trip takes, multiplied by the speed of sound in water.

Single beam sonars enabled faster and more accurate mapping of the bottom of the ocean than lead lines, but they shared some of the disadvantages of their predecessors. As the name suggests, they could only send a

*This term has nothing to do with sounds that we can hear; instead, it refers to the geographical name of a narrow body of water joining two larger bodies of water—a *sound*, also known as a *strait*. We'll see later, however, that associating depth sounding with the most common meaning of the word *sound* isn't too far off.

†Traditionally, water depth has been measured in fathoms, each equal to six feet. In the 1850s, navigators sounding the Mississippi River would yell "by the mark twain!" to mark a depth of two fathoms. One of the navigators happened to be one Samuel Langhorne Clemens, who would come to be better known as Mark Twain.

single ping at a time, so mapping larger areas—the oceans cover more than 70 percent of our planet, after all!—was still extremely slow. This prompted another major improvement: multibeam sonars, developed in the 1960s.

Mowing the Ocean

Instead of a single ping, a multibeam sonar sends a whole swath of signals, allowing researchers to map a belt of the seafloor in one go. Today's multibeam sonars can cover a width of up to a few kilometers at once, although the quality of measurements drops off farther from the ship, where the beams travel a long distance to the seafloor. Scientists often refer to the process of mapping with multibeam sonars as "mowing the lawn" because the ship moves back and forth in straight lines to cover the whole area.

The mechanism of sonars is simple, but their correct use requires a lot of care. First and foremost, the correct depth depends on two measurements: the time of the signal's round trip and the speed of sound. The sonar deals with the former, but the researchers must ensure they have a precise value for the latter. In general, sound travels at around 1,500 meters per second in water—about 4.5 times as fast as in the air*—but the speed changes with salinity, temperature, and pressure. Other important factors are the frequency of signals sent and the speed of the ship, which together determine the level of detail included in the map. The faster the ship is moving and the less frequently the sonar pings, the fewer signals are received per area covered. In deeper parts of the ocean, either the ship must slow down or the pinging rate must increase to compensate for the longer time the signals need to return to the ship.

The map's resolution also depends on the depth of the water, because each swath contains the same number of beams but covers a different area. In shallow waters, all measurements fit in a small area, which lets us see many details. In deep waters, the area covered by a swath is larger but, since the measurements are more spread out, the resolution is lower. For example, in water 100 meters deep, the multibeam sonar might cover a belt about 500 meters wide, taking measurements about 1 meter apart, while at a depth of 3,000 meters, the same sonar will take measurements about 10 meters apart along a width of about 9,000 meters. It's much like holding a flashlight

* In water, particles are packed much closer than in the air, which allows the sound wave to move faster.

in a dark room. If we aim the light directly at the floor, we see objects in the small circular beam clearly. On the other hand, if we angle it out, the light will encapsulate a larger oval area, but the beam will be weaker—and the farther away from the floor we hold the flashlight, the weaker it will be. With multibeam sonars, it's always a balance between the area covered—and, therefore, the speed of mapping—and the resolution. In particular, because the signal has a longer distance to travel on the outer edges of the swath of beams, in deep areas of the ocean researchers often reduce the maximum angle of the swath to increase the resolution.

In addition to the depth, the sonar measures the strength of the signal upon its return, which informs us about the type of seafloor, called backscatter. In general, rocky surfaces will reflect a stronger signal than muddy surfaces, and a smoother seafloor will reflect more sound than a bumpier seafloor.* The combination of two types of data—the time it takes the signal to return to the sonar as well as its strength upon return—lets researchers create detailed maps of the bottom of the ocean.

Aboard the *Nautilus*

Curious about what it's like to participate in mapping expeditions, I contact the Ocean Exploration Trust (OET), a nonprofit organization that coordinates research expeditions onboard a specialized ship, *Nautilus.* Oceanographer Derek Sowers, who has spent many months at sea, kindly allows me to interrupt his preparations for an upcoming expedition with dozens of questions. An oceanographer by training and nature lover at heart, about a decade after joining OET he still gets a thrill out of boarding *Nautilus.* "You get to go to places that are very poorly understood," he tells me with a wide grin. "Map them for the first time, see what's there, make discoveries and share it with a lot of people—so it's pretty amazing." He and the team go to what seems like the middle of nowhere, even though they tend to focus on the US exclusive economic zone.† Beyond the waters around the mainland United States, this zone contains vast areas surrounding remote islands in the Pacific, including the US's largest protected

* For the same reason, in empty rooms sounds echo much more than in rooms full of soft sofas, blankets, and cushions.

† For a reminder of what this zone is, see the footnote on p. 87.

marine area, with the beautiful name Papahānaumokuākea Marine National Monument.

OET runs a few expeditions every year and, while each lasts about a month, the research team starts preparations way in advance. To plan the survey, they gather all possible information about the area, such as data from previous expeditions or satellite images. Even low-resolution images can indicate, for example, a large sea mountain that they might want to map in detail when they get there. A vague idea about the depth of the seafloor allows researchers to choose appropriate sonar settings, such as the width of signal swaths. When they finally board the ship, the work goes on 24/7—time on the water is too expensive to be wasted, and the Seabed 2030 clock is ticking. A few dozen scientists onboard rotate sleep schedules so that everything runs smoothly night and day. By the end of the expedition—and they have a hard stop determined by the amount of food the ship can carry—they will hopefully have gathered and processed new, interesting data, which will soon become publicly available. In the end, that's their main goal: to create a comprehensive database for fellow researchers as well as for the public.

Mappers of the ocean floor stand out among researchers in most other disciplines, where research still tends to happen behind closed lab doors. Scientists engaged in Seabed 2030 focus on documenting the data to make them easy to use—which, unfortunately, is still rare—saving their peers time and reducing the likelihood of duplicated effort. They also educate the public about our rich, beautiful oceans, following the belief that we care for what we know, and ignorance leads to neglect. And, in this way, they continue the legacy of a trailblazing mapper of the seabed by the name of Marie Tharp.

Is This a Man's World?

In the aftermath of the Japanese attack on Pearl Harbor in 1941, as young men were flocking to join the US military, emptying campuses all over the country, some university departments opened their doors to female students. This expanded the options available to women, which had so far been limited mostly to becoming a teacher, a secretary, or a nurse—none of which appealed to Marie Tharp, then a twenty-one-year-old student at the University of Ohio. Having changed her major every semester, she ended up graduating with majors in English and music and, despite pursuing four minors in addition, she still hadn't found her calling. "Girls were needed to fill the jobs left open because the guys were off fighting," she later recalled, and this new demand allowed her to earn two degrees: a master's in geology from the University of Michigan and a bachelor's degree in mathematics from the University of Tulsa.

A few years later, when she was looking for a job in New York City, the two latter degrees outweighed her biggest flaw—that is, being a woman—and she was offered a position as a draftsperson within Columbia University's geology department. The position was well below her qualifications but, in the mid-twentieth century, a female graduate couldn't count on more than assisting male students working toward their first degree.

One of the first women at the newly created Lamont Geological Observatory (today called Lamont-Doherty Earth Observatory), for decades Tharp analyzed the data gathered by her younger colleague, a geologist named Bruce Heezen. Men wanted to do the exciting fieldwork, not the dull—in their opinion—data wrangling, which they left to their female colleagues. Tharp couldn't board the research ships that the institution regularly sent to the ocean to record observations—again, simply because she was a woman—so she stayed in the office, exploring the numbers. But, while sitting in her office chair, Tharp ended up discovering more than any of the seafaring men.

A Tectonic Discovery

Heezen would regularly supply Tharp with long paper scrolls containing sonar readings of the ocean's depth along a research ship's route. These data, represented with a wiggly line, were hard to interpret, which prompted Tharp to design a new way of visualizing the seafloor. First, she'd plot the ship's track on a map, to locate it geographically. Then, she'd scale up the lines of sea depths, marking deep points at the bottom of the diagram and shallow points close to the top of the diagram, thus visualizing an ocean floor profile. This way, the sometimes subtle differences in depths became clearly visible. Finally, she'd redraw these exaggerated depth profiles along the ship tracks drafted in the first step. As a result, Tharp would get a map of the ocean with depth readings, putting the data into a geographical context. The data, visualized this way, revealed something that would revolutionize science.

In the 1950s, it was generally assumed that the ocean floor was featureless: it was believed that mountains, hills, and valleys belonged to the dry land, but the bottom of the ocean was essentially flat. That was not, however, what Tharp saw after drafting six west–east profiles of the North Atlantic that corresponded to six different, approximately parallel ship tracks. While not identical, they all featured a V-shaped feature around the middle. Tharp suggested that it might have been a rift valley dividing a massive mountain range extending in the north–south direction along the Atlantic, which Heezen disregarded as "girl talk" and told her to redo all the work. Why did Tharp's innocent observation seem so outrageous to her colleague?

Today, we know that the Earth's outermost layer, called the lithosphere, consists of tectonic plates slowly moving toward or away from each other. As the plates interact, coming together or pulling apart, they create new landforms, which can be a process as slow as mountain building or as rapid as earthquakes. We have a lot of evidence that millions of years ago, the continents didn't look anything like they do today, and they're always slowly moving, even as I'm typing these words.

The theory of plate tectonics, which explains phenomena like earthquakes, volcanism, and the formation of mountains, is now generally accepted and taught at schools. But as late as the 1950s, believing in plate tectonics or the related theory of continental drift was akin to admitting to being a flat-earther.* Heezen knew that a rift valley would have been created by tectonic plates moving apart, so accepting Tharp's discovery would label him a "drifter." This could ruin his career in a world of "fixists," who explained the formation of mountains through the contraction of the Earth's surface and not the collision of tectonic plates. Despite knowing that continental drift was considered "almost a form of scientific heresy," Tharp was confident in her insights, so she kept plotting any ocean depth data to which she had access. As the rift valley appeared over and over, Heezen couldn't deny any longer that his assistant was on to something. Now they had to persuade the rest of the scientific world.

*Continental drift suggests that the Earth's continents have been moving with respect to one another. Plate tectonics describes *why* it's happening: the continents move because the Earth's lithosphere is divided into moving and interacting plates.

Seeing Is Believing

Tharp knew that no text would convince a skeptical scientist of the existence of the oceanic rift valley unless it was accompanied by a map, because, as she famously said herself, "a picture is worth a thousand words" and "seeing is believing." Due to the ongoing Cold War, however, the US Navy had classified any detailed contour maps of the seafloor. The ban didn't mention less precise physiographic diagrams, which present the bird's-eye view of the Earth's surface features. A physiographic map of the ocean floor would not only circumvent the navy's restriction, Tharp thought, but also visualize the varieties of seabed texture, from the smooth abyssal plains to the rugged mountains. And, most importantly, it would allow even nonspecialists to imagine what the ocean floor looked like, "as it could be seen if all the water were drained away."

Tharp and Heezen didn't invent physiographic maps, but until that point, these visualizations had always represented landforms on continents. Creating the first detailed diagram of the bottom of the Atlantic Ocean posed a formidable challenge, especially given the data scarcity. Lamont's ships were equipped with modern sonars able to measure the ocean depth continuously as long as there was enough electric power, which would go off every time the ship's fridge would open. In these moments, as Tharp recalls, the recorded depths would be "as bottomless as the crew's appetite." Even with perfect equipment, however, one ship was able to record only one depth profile—a single line in a vast ocean—and, just as today, the number of expeditions was limited by time and money. So, Tharp and Heezen had to get creative about filling in the gaps: they'd look for data from alternative sources, they'd make educated guesses about depths at unmeasured points, and the areas of the map where they had no clue about the ocean's depth they simply covered with a large legend, "like the cartographers of old."* All these efforts paid off: by 1956, their first physiographic map of the North Atlantic was ready.

* Tharp admitted to having suggested including images of mermaids and shipwrecks to cover these blank spots but, apparently, Heezen wasn't enthusiastic about the idea.

Shaking the Foundations of Geology

While Tharp was tinkering with her revolutionary map, Bell Telephone Laboratories (BTL)—the famous research company, home to inventions from the transistor to the laser to multiple programming languages—asked Lamont to help them figure out safe locations for new telephone cables. Even today, phone calls and the internet work thanks to a massive network of underwater cables connecting users across continents. We appreciate this system only when it fails, be it due to human error or a natural disaster like 2012's Hurricane Sandy, which caused major internet disruptions on the US East Coast. BTL had understood the vulnerability of underwater cables since at least the 1920s, when currents caused by earthquakes had destroyed some of their transatlantic cables. To prevent similar situations in the future, they needed a map of past earthquake locations which they could then avoid encroaching on.

Heezen hired an artist, Howard Foster, to meticulously plot tens of thousands of such locations by hand because computers weren't available back then. When Tharp compared her map with Foster's creation, drawn on the same scale, she did a double take: she had discovered that earthquakes clustered in the rift valley. This provided a strong piece of evidence for the continental drift theory: tectonic plates moving away from each other created both earthquakes and the cleft in the ocean floor.

Both Tharp and Foster extended their maps to other parts of the world, using data from expeditions to the Indian Ocean, the Pacific, and other seas. They kept observing the same pattern: all earthquakes happened around mid-ocean ridges, and all mid-ocean ridges were seismically active. What's more, the rift in the Indian Ocean was continued on land as the East African Rift Valley. This led to an earth-shattering, if I may, conclusion: the rifts made up a worldwide connected system.

The reactions of Heezen and Tharp's fellow scientists to the revelation varied: some considered them geniuses, others lunatics. A renowned geologist from Princeton University, Henry Hess, was impressed. "Young man, you have shaken the foundations of geology!" he exclaimed after Heezen's talk about the global system of rift valleys. On the other hand, an ocean explorer, Jacques Cousteau, went as far as crossing the Atlantic Ocean with a camera lowered to the seafloor in an attempt to disprove the existence of rift valleys. He expected to find nothing but instead returned with movies showing the beautiful rift valley, which proved invaluable in persuading other scientists and the public that Lamont's discovery was true.

Tharp and Heezen's innovative maps made it easy to understand what the ocean floor looked like and were quickly noticed by *National Geographic.* The magazine asked them to collaborate on illustrating an article about an expedition to the Indian Ocean with a talented Austrian artist, Heinrich Berann, known for his stunning maps of mountainous landscapes, often used to promote ski resorts.* His techniques worked equally well for underwater mountain ranges, which doesn't mean the process was either quick or painless. For three years, Tharp and Heezen regularly traveled between New York and Innsbruck, delivering to Berann any new data they had managed to process and taking early drafts of the map back home. The effort was worth it: even today, many of us recognize the beautiful, poster-sized maps of the ocean floor published by *National Geographic* in the 1960s and 1970s, which brought Tharp and Heezen's discoveries to thousands of magazine subscribers.

"I think our maps contributed to a revolution in geological thinking, which in some ways compares to the Copernican revolution," Tharp recalled years later, and she wasn't boasting. In 1912, a German meteorologist and geophysicist by the name of Alfred Wegener had already proposed the theory of continental drift, but nobody had taken him seriously until decades later when Tharp's novel representations of ocean depth data revealed the rifts at the bottom of our oceans. The work of Lamont's scientists proved Wegener's theory and helped explain that continents drift due to moving tectonic plates.

Today, it's amazing to think that continental drift and plate tectonics—theories we consider as obvious as evolution, gravitation, or the germ theory of disease—entered textbooks only in the 1970s or even the 1980s in some countries. This makes my parents the first generation to study the dynamics of the Earth's upper layer at school, and who knows if they'd have learned about it if Marie Tharp hadn't created her revolutionary map. Mapping the ocean floor is so difficult that at the time of writing, we're not even close to completing the process. Which begs the question: If the Earth's

*Berann partially owes his employment with *National Geographic* to his young daughter, who had written to the magazine. "I've been looking at your maps and my father can paint better than you can," she said. When their chief topographer visited the talented artist in Innsbruck, Austria, it turned out she was right.

upper layer is so hard to study, how can we hope to know anything about its deeper parts?

Mapping the Earth's Interior

Until recently, studying the Earth's interior was all but impossible. To glean information, we've tried digging holes—with equipment as simple as shovels and as complex as the research vessel *JOIDES Resolution,* which drills into the ocean floor and collects samples from the Earth's inside—but even with the availability of extraordinary pieces of machinery we haven't yet been able to get very deep. Instead, we study the chemical composition of rocks on and close to the surface—such as the molten rock called lava that erupts from a volcano or a fissure onto the Earth's surface—to learn about the conditions inside the Earth, where these rocks formed. Satellite missions help too, by measuring and mapping the Earth's gravitational field: stronger gravity indicates larger mass and density within the Earth's interior, so variations in the gravitational field suggest what materials different parts of the Earth consist of. But possibly the most ingenious way of mapping the Earth's interior harnesses the power of earthquakes.

I'm lucky to have grown up in a country that barely experiences earthquakes, but billions of people live on a ticking bomb. As Marie Tharp figured out, nobody living around a tectonic plate boundary can feel safe; indeed, earthquakes endanger countries as diverse as Chile, Japan, and Turkey.

In often tragic news reports about these natural disasters, we hear about the epicenter, which is the point on the Earth's surface directly above the place where the earthquake originated. As the ground shakes, forcing locals to scramble for shelter, seismic waves travel through the Earth. These waves are caused by sudden movements within a planet, for example, an earthquake, a volcanic eruption, or an explosion. They move through the Earth's interior and surface, carrying acoustic energy. By recording the time it takes for them to reach other parts of the world, as well as the amount of energy they carry, we can learn a lot about the Earth's interior.

The first detector of earthquakes was invented in the second century CE by a Chinese scientist named Chang Heng. This first seismoscope consisted of an urn surrounded by eight dragon heads, each holding a ball in its mouth, perched above a model of an open-mouthed toad. During an earthquake, some of the dragons would drop their ball into the open mouth of

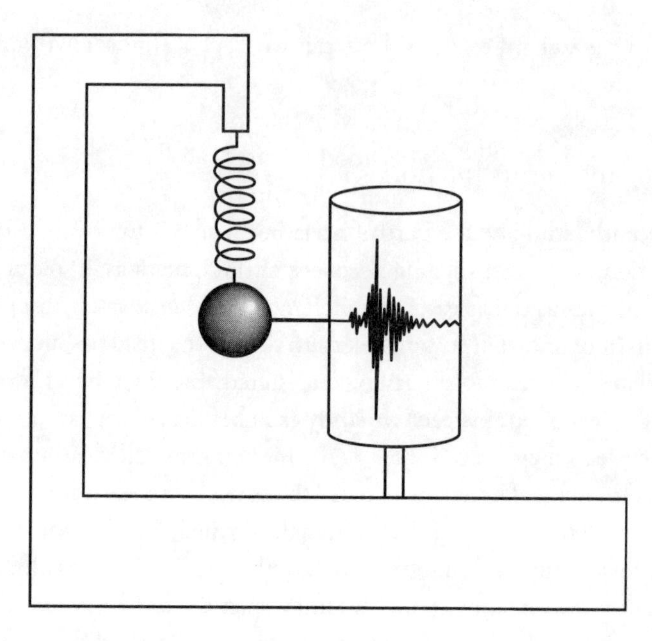

Figure 8.1 During an earthquake, the frame moves with respect to the suspended mass, which is recorded as wiggles on a rotating paper scroll.

the toad sitting below it, indicating not only the occurrence of seismic waves but also the rough direction from which they were coming. Over centuries, the design of instruments able to measure the time and intensity of an earthquake—called seismometers or seismographs*—has improved. A simple seismometer consists of a frame firmly attached to the ground and a heavy mass suspended from the frame on a spring, which together form a pendulum. As the Earth shakes, the frame follows the ground's movements, but the heavy mass stays in place due to the spring absorbing the shocks. Early seismographs also contained a rotating paper scroll attached to the frame and a pen attached to the hanging mass. This way, as the frame moved with respect to the pendulum, the pen would create wiggles on the paper, recording the shocks. Today, the pen and paper are replaced by a digital output, but the pendulum-based mechanism has stayed the same. Such outputs, collected from many seismometers, helped create a map of the Earth's interior.

*Strictly speaking, seismographs consist of a measuring device—the seismometer—and a recording device. However, these two terms are usually treated as synonyms.

Catching the Waves

Almost ironically, ancient scientists were able to accurately map remote stars and planets, but we hadn't understood much about what's underneath our feet until the late nineteenth century. Born in India in 1858, Richard Dixon Oldham followed in the footsteps of his father, a geology professor interested in Himalayan earthquakes. His discovery in the aftermath of a large earthquake that hit Assam in 1897 forever changed the way we study the Earth's interior. After this event, which took the lives of over 1,500 people, Oldham noticed that European seismometers registered seismic waves of three different shapes, thus identifying three types of seismic waves: surface waves, P waves and S waves.

P waves stand for primary or pressure waves, which, as the name suggests, travel faster than other types and thus arrive first at the seismic station. P waves are longitudinal, which means that they move the Earth in the direction they propagate, like a Slinky spring toy pushed forward. The arrival of a P wave is soon followed by a slower S wave, known as a secondary or shear wave. S waves are transverse, so they move the Earth in the direction perpendicular to their motion, like a Slinky moving up and down. This type of movement requires a rigid material, which means that as opposed to P waves, S waves can only travel through solids. This fact comes in useful in mapping the Earth's interior.

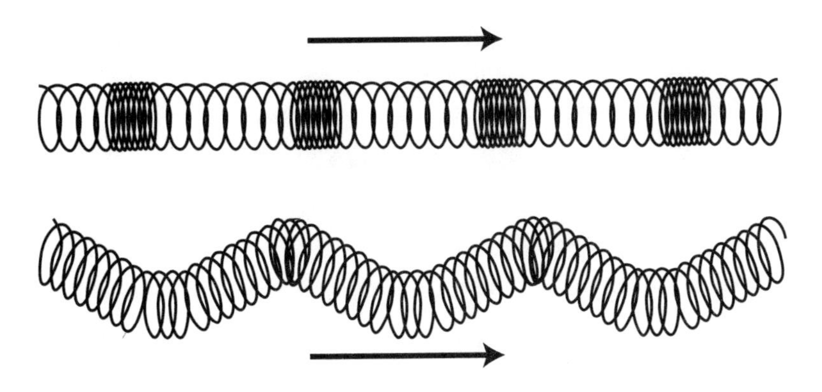

Figure 8.2 P waves (*above*) displace the ground in the direction of their movement. S waves (*below*) displace the ground in the direction perpendicular to their movement.

How Waves Bend

Being able to use our eyes almost every waking moment doesn't, unfortunately, make us experts in processing light waves, which still manage to trick us. If you put a spoon in a glass of water, for example, it will appear to bend. This is due to the refraction of light, which moves at different speeds in air and water. As it passes from one medium to another, light enters at one angle but emerges at a different angle, causing peculiar optical illusions—like the apparent bending of the spoon. The relationship between these angles and the speeds of light in the two media is captured by Snell's law.* Instead of looking at the formula, let's try to understand where the law comes from.†

A beam of light always chooses the fastest route between two points. In one medium, the fastest route would simply be a straight line, but it's not the case when light travels at different speeds at different parts of the journey. To visualize that, imagine that you're enjoying a sunny day on a beach when you spot someone drowning in the lake. Of course, this person needs urgent help, so—like the beam of light—you need to get there as soon as possible. Intuitively, you'd want to cover a bigger part of the distance running along the beach to minimize the swimming time, since it's likely that you swim more slowly than you run (even if you're Michael Phelps). A light wave does the same: it's slower in water, so to minimize its total journey time, it needs to spend less time in the water than in the air. This causes it to bend, or refract.

Snell's law applies not only to light but to all kinds of waves. Like air and water, different layers of the Earth's interior are made of different materials, creating a boundary that refracts seismic waves. So, a map of the Earth's interior that included the thickness of layers and the speeds of seismic waves in different media, together with Snell's law, enabled scientists to compare a model of seismic wave propagation with reality.

*Snell's law is named after Willebrord Snell, whom we met when we discussed triangulation, although a Persian scientist named Ibn Sahl had already formulated this law in the tenth century.

†If you're curious, here it is: $\sin(\theta_1)/\sin(\theta_2) = v_1/v_2$, where v_1 and v_2 are the speeds of light in the two media, and θ_1 and θ_2 correspond to the angles in Figure 8.3.

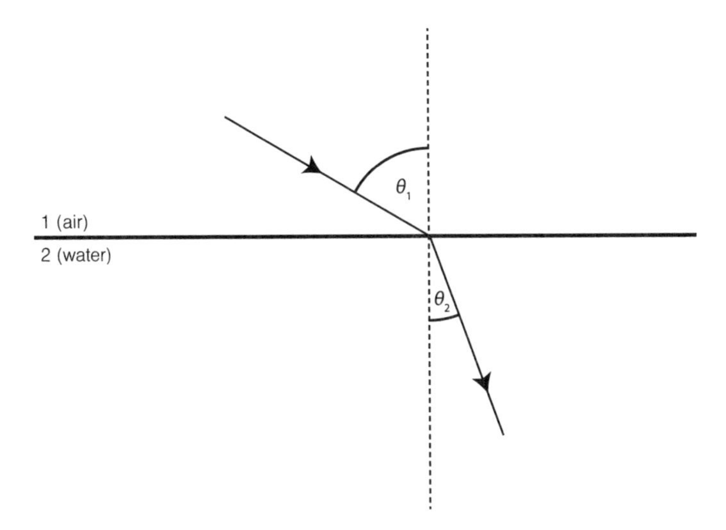

Figure 8.3 Light waves are slower in water than in the air, so to reduce the total journey time, they bend as they travel from one medium to the other.

In the Shadow

If in the whole Earth's interior, the conditions—for example, the pressure and the rock composition—were identical, seismic waves generated by an earthquake would propagate in slightly curved paths to seismic stations all over the world. In the early 1900s, scientists realized that this wasn't the case. By analyzing the data from the expanding global network of observatories, they noticed a peculiar pattern: seismographs close to the epicenter registered S waves, but about 103 degrees away from the epicenter, the signals would suddenly stop, as in Figure 8.4. In a homogenous Earth, one would expect a gradual decrease of the signal's strength due to a larger distance covered, so the abrupt lack of signal didn't comply with this model. Something fishy was happening deep in the Earth! Because S waves propagate only through solids, there was only one explanation for the so-called S wave shadow zone: the center of the planet must contain a liquid layer that stops S waves.

The existence of a liquid core, as the innermost layer of the Earth is called today, was also suggested by the patterns in the arrivals of P waves. These patterns were even more curious: like S waves, they would arrive at stations up to about 103 degrees from the epicenter and then they'd stop, only to re-emerge at about 150 degrees. In other words, P waves would reach

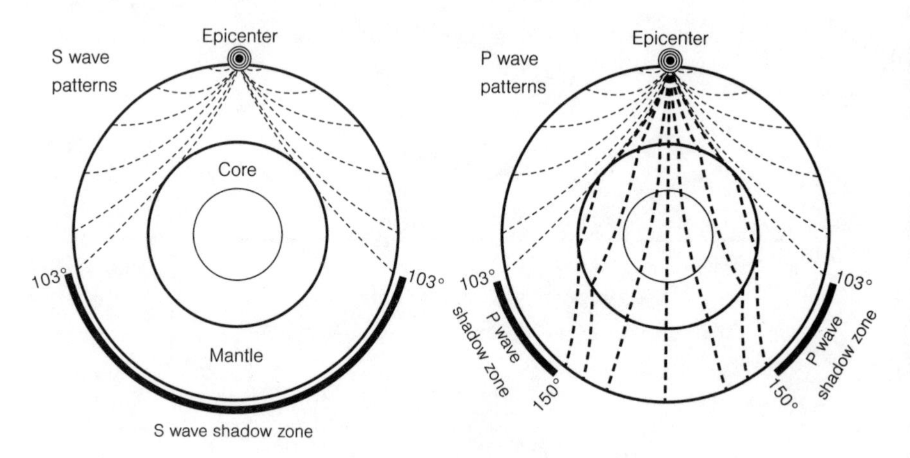

Figure 8.4 S waves aren't detected beyond about 103 degrees on either side of the epicenter. P waves aren't detected between about 103 and about 150 degrees on either side of the epicenter.

seismographs close to the epicenter as well as those on the other side of the Earth, but not those in two relatively narrow areas called P wave shadow zones. Moreover, in the antipodal area of the epicenter—that is, in the region beyond the P wave shadow zone—the waves would arrive later than a homogenous Earth model predicted. Since waves move slower through liquids than solids, a liquid core would explain this pattern.

Building a model of the Earth's interior involved dozens of scientists publishing one paper after another, gradually refining their ideas. Over a few decades, this map evolved from a simple theoretical concept to a model supported by seismic evidence and mathematical computations. It was widely accepted until the 1930s, when a Danish Earth scientist named Inge Lehmann took a deeper look—quite literally.

Never Too Late to Study Earthquakes

Having learned about the obstacles Marie Tharp had to face simply to do her work in the 1940s, I expected that a woman born in 1888 stood no chance of a scientific career. And yet, Inge Lehmann did manage to gain the respect and recognition of her male colleagues.

Inge's parents gave her the best start possible in life, sending her to the first Danish co-educational school, a progressive institution run by Hanna Adler, an aunt of the famous physicist and Nobel Prize laureate Niels

Bohr.* One of the first women to graduate from Copenhagen University, the headmistress was adamant about treating girls and boys equally. Boys would cook and sew, girls would run around with a football—and every child would study science. Used to having the same rights as boys, Lehmann would feel "some disappointment later in life when [she] had to realise that this was not the popular opinion."

Lehmann went on to study mathematics, physics, chemistry, and astronomy at the University of Copenhagen. She also had the chance to visit Cambridge's Newnham College for a year, where she was unpleasantly surprised by "the severe restrictions inflicted on the conduct of young girls, restrictions completely foreign to a girl who had moved freely amongst boys and young men at home." Alas, her fierce ambition got the best of her, and she was forced to take a break from studies due to severe fatigue. While gaining back the strength to return to university, she worked in an actuary's office, where she learned the computational skills indispensable in her future scientific career.† By 1918, she was ready to return to the University of Copenhagen, from which she graduated two years later.

After graduation, Lehmann continued actuarial work until 1925, when her career trajectory took a sharp turn. She became an assistant to a researcher responsible for planning and installing a network of Danish seismic stations and, in this way, thirty-seven-year-old Inge Lehmann became a seismologist—a scientist studying earthquakes and seismic waves. She soon passed all necessary exams and became the chief of the brand-new Royal Danish Geodetic Institute's seismological department. I can't think of a better proof of a woman's skills than being trusted to manage her male colleagues.

Lehmann's job description didn't involve research per se; instead, she was supposed to interpret seismic data and publish the measurements. This task gave her an overview of the quality of seismic data from different observatories, which inspired her attempts to standardize and improve seismograph readings. This was relatively easy to do in accessible stations like Copenhagen. More remote stations, for example Scoresby Sund, a desolate station in Greenland connected to the world only by the annual arrival of a

*Like many Jews during the Second World War, eighty-four-year-old Hanna Adler was detained in Horserød camp, a de facto concentration camp. She regained her freedom after a passionate letter from 400 of her grateful former students, including Inge Lehmann.

†Actuaries use advanced mathematics and statistics to assess and manage risk and uncertainty. They often work in the insurance industry.

boat, posed a major challenge. It was this behind-the-scenes work that gave Lehmann and her peers good quality data crucial to improving the map of the Earth's interior.

A Solid Core

By the early twentieth century, scientists correctly assumed that the Earth was made of layers, like onions.* But at the time they incorrectly assumed the existence of only three main layers: the core in the middle, then the mantle and the crust.

In 1929, to her surprise, Inge Lehmann found out that European observatories had detected seismic waves from a large earthquake in New Zealand. Given that Europe was in the P wave shadow zone of the epicenter, no waves should have appeared there. In comparison to her predecessors in the field, Lehmann had access to a larger network of more sensitive seismometers. She also was quite lucky (which can't be said about the New Zealanders who suffered in the earthquake), because the antipode of the epicenter lay in Portugal, an area dense with seismometers. Had the earthquake happened in Italy, for example, or India, or California, the P wave shadow zone would have mostly consisted of oceans. It meant that it wasn't a single, potentially broken seismometer that detected P waves where there should have been none; instead, with observations from multiple seismic stations at her disposal, Lehmann was sure that there must have been a different reason. Having considered and dismissed other possible explanations for P waves violating the shadow zone, she concluded that the widely accepted map of the Earth's interior must have been wrong. An additional layer was needed.

Lehmann wondered how seismic waves would behave if the Earth's core had two layers: a solid inner core and a liquid outer core. The strength of her reasoning lay in the appreciation of simplicity: as opposed to her colleagues, she stripped the Earth's model down to all but the crucial assumptions. For example, while Oldham and others struggled with the complex math of curved paths of seismic waves, Lehmann assumed that seismic waves traveled in straight lines, which allowed her to apply simple trigonometry. She also did away with varying wave speeds and assigned a

*Or ogres. I'm so excited to finally have an opportunity to drop a *Shrek* reference.

constant velocity within each layer. These were significant simplifications, but they allowed her to see the bigger picture.

Lehmann applied Snell's law and trigonometry to estimate the paths of seismic waves in the model with the double-layered core. Lo and behold, this new model of the Earth's interior explained the real-world data from seismic stations. In 1936, Lehmann published her literally groundbreaking results in a paper entitled simply "P′" (pronounced "P prime"), which stands for P waves that pass through the mantle, the core, and then the mantle again. The tone of the paper was conservative. Lehmann didn't claim to have established what the Earth's interior looked like, only that her new model agreed with the observations. Having studied the quality of seismic observations, she was well aware of flaws in the data she had used to validate her model, so she hoped the article would encourage further investigations rather than be the definitive word. In a world that praises overconfidence and condemns nuance, it must have taken a lot of bravery to include those caveats. Even today, it's rare to find a scientific paper as open about the limitations of the authors' research as it is about its contributions. It might be due to Lehmann's honesty and modesty that, after the publication of "P′," the map of the Earth's interior was quickly filled in with more detail—and we keep improving it to this day.

Across the Pond

Not long after Lehmann's seminal publication, the Second World War broke out, making seismic observations hard to continue. The United States took control of Greenland's seismological stations, on which Lehmann and others had relied, and the crucial international exchange of data became all but impossible. After the war, however, seismologists became invaluable—not necessarily because they could track earthquakes, but because they could detect underground nuclear explosions. This attracted additional funding for the scientists and laboratories in the field, which helped further Lehmann's career.

Lehmann received an unexpected visit from Maurice Ewing, the director of the Lamont Geological Observatory—the same institution where Marie Tharp spent her whole career. Impressed by Lehmann's work, he invited her to help them study a newly discovered type of surface seismic wave. Soon, she was on her way to the United States, carrying a bunch of seismograms to compare with Lamont's data. She was treated as a celebrity there, and Ewing made sure she had everything she needed to be comfortable and productive.

Back in Denmark, however, she was refused a position as geophysics professor at the University of Copenhagen, which prompted her early retirement in 1953. Freed from other responsibilities, Lehmann committed the rest of her long life to geophysical research, rich in international collaboration. Most of her fifty-eight publications appeared after her retirement, with the last one written at the age of ninety-nine—five years before her death in 1993.

From Oatmeal Boxes to Tomography

Inge Lehmann discovered the Earth's inner core without using electronic computers. She collected, organized, and analyzed all the data by hand. That's how Nils Groes recalled his relative's research process:

> I remember Inge one Sunday in her beloved garden on Søbakkevej; it was in the summer, and she sat on the lawn at a big table, filled with cardboard oatmeal boxes. In the boxes were cardboard cards with information on earthquakes and the times for their registration all over the world. This was before computer processing was available, but the system was the same. With her cardboard cards and her oatmeal boxes, Inge registered the velocity of propagation of the earthquake to all parts of the globe. By means of this information, she deduced new theories of the inner parts of the Earth.

To learn how we map the Earth's interior almost a century after Lehmann's monumental discovery, I speak with seismologist Paula Koelemeijer of the University of Oxford (and her adorable four-month-old, who insists on participating in the conversation). Koelemeijer's research focuses on the Core–Mantle Boundary, but she's interested in interdisciplinary topics ranging from Mars's interior to the impact of COVID-19 lockdowns on anthropogenic seismic noise to the seismic vibrations of elephants. She truly knows what's hot in the seismology world!

Koelemeijer tells me that for decades following the discovery of the Earth's inner core, seismologists used the same method as Lehmann—

combining arrival times from seismic waves around the globe—to add more details to the basic inner core–outer core–mantle–crust model. By splitting the main, uniform layers into thinner ones, they accurately described the Earth's structure along one dimension, creating a map of the onion-like Earth. According to Koelemeijer, current predictions of the time it takes a seismic wave to travel from one side of the Earth vary from the actual arrival times by only a few seconds—which, given that the wave's journey lasts about twenty minutes, is an impressive accuracy. No matter how many layers we assume, however, these models reduce our three-dimensional planet to only one dimension and don't enable us to map speed and density variations *within* the layers.

In the late 1970s, seismologists started looking at the three-dimensional structure of the Earth's interior, creating today's state-of-the-art way of imaging it: seismic tomography. From the Greek *tomos* ("slice, section") and *graphō* ("to write, to describe"), its name indicates that it allows researchers to create three-dimensional images of the Earth and slice through them to look at two-dimensional cross-sections. As with medical tomography (CT scans), which uses differences in energy transmitted by X-rays within the body to create three-dimensional images of internal organs, seismic tomography uses differences in speeds of seismic waves traveling through the Earth to create three-dimensional images of its interior. While Lehmann and her peers could only model a limited number of homogenous layers, today's seismologists can look at the Earth's structure in as much detail as they need to.

It all starts with a one-dimensional model that assumes constant velocities in each layer. In reality, the velocities within a layer vary, so we can think of the constant velocity as an average over these variations. If a seismic wave arrives from an earthquake at a seismic station earlier or later than the model predicts, we know that it passed through a region whose properties differ from the average—but where exactly did it happen? To answer that, we need another earthquake and another seismic station so that the paths of the two waves cross. If the second wave's arrival time differs from the model in a similar way to the first wave's arrival time, chances are that the variation happened at the crossing, as Figure 8.5 shows. Seismologists analyze millions of such crossing paths, which allows them to build a huge system of equations. The solution to this system translates to a three-dimensional image of the Earth's interior that includes such speed variations within the layers.

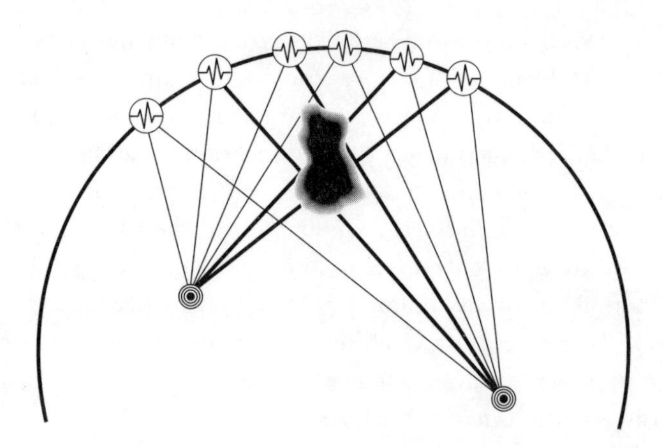

Figure 8.5 Observatories register a late arrival of seismic waves that pass through a low-velocity structure. By understanding where their paths crossed, researchers can find out where the velocity was lower than average.

Blobs in the Earth

Thanks to seismic tomography, researchers can capture more details than Lehmann ever could. Today we know that the Earth's layers aren't perfect rings with uniform properties, which is what Lehmann had to assume to compute anything by hand. While the general layered model is correct, the layers contain many irregularities. To capture these varying properties, however, seismologists need powerful computers—no human would be able to process the data from millions of seismometer outputs corresponding to thousands of earthquakes. The developments in computing allowed researchers to use more data and reduce the number of approximations needed. Even so, some approximations are still necessary, which means that we don't end up with *the* map of the Earth's interior, but a collection of maps resulting from different choices made by different seismologists. All these maps capture the same large features, but they differ on smaller structures within the Earth.

Koelemeijer points out that one-dimensional models, mostly developed before the 1980s, are already quite accurate. The variations from the average velocities detected thanks to seismic tomography are on the order of only about a few percent—but this was enough to change the map of the Earth's interior. In Koelemeijer's opinion, the most important thing we've discovered thanks to three-dimensional imaging is the evidence for heat exchange between the upper mantle and the lower mantle. In the late twentieth century, geophysicists couldn't agree on whether the mantle consisted of two separate layers, or if the mantle has one heat budget. This might seem like a purely theoretical debate, but solving this conundrum allowed us to better understand how quickly the Earth has cooled over time. Three-dimensional images revealed low-velocity structures penetrating through the layers called mantle plumes (or, more informally, blobs). These columns of hot material might be responsible for the volcanic activity outside of tectonic plate boundaries. As Marie Tharp discovered, volcanoes appear where tectonic plates meet, but this doesn't explain regions such as Yellowstone, Hawaii, or Iceland, which are nowhere near such a boundary. Now scientists suspect that these anomalous volcanic areas are fueled by mantle plumes.

In addition to mantle plumes, seismic tomography revealed two continent-sized structures on the core-mantle boundary called large low-velocity provinces (LLVPs). They're so large that together they make up about one-fourth of the core's surface area. It's still unclear how LLVPs relate to mantle plumes: some say that mantle plumes come from these two large structures, while others claim that LLVPs themselves are clusters of multiple mantle plumes. Seismologist Sanne Cottaar, the leader of the Deep Earth Seismology group at the University of Cambridge, tells me that the origin of mantle plumes and LLVPs is one of the biggest unanswered questions in seismology. What are they? Have they formed together with the Earth or have they grown over time? As we fill in the map of the Earth's interior, new questions arise.

Finding the Epicenter

Since mapping the Earth's interior involves studying paths of seismic waves from the epicenter to the seismograph, we wouldn't get far without the earthquake's precise location. At the beginning of her seismological career, Inge Lehmann attended a conference in Prague, where she witnessed endless disagreements about the precise paths of seismic waves. She figured—and confirmed with her further research—that a big component of discrepancies between seismologists' results was the inaccurate determination of the epicenter.

The epicenter is relatively easy to estimate when an earthquake happens close to a populated area, but what if it's the ocean floor that shakes? We can infer the earthquake's location using observations from multiple seismic stations and some math. To simplify the explanation, we'll assume a shallow earthquake, whose exact location in the Earth's interior—called the focus—and its epicenter are approximately the same.*

As we've seen, the S wave arrives at a seismic station with a delay with respect to the P wave. We know how fast P and S waves travel through the Earth, which gives us enough information to calculate the distance to the epicenter based on the speed = distance / time formula.† This only lets us draw a circle of radius equal to this distance around the seismic station; to establish the exact epicenter, we need to repeat the exercise for three observatories, which is how this process got its name: *trilateration.* More generally, trilateration means finding the location of a point in a three-dimensional space given its distances to three other points.‡ It is also the basis of the global positioning system (GPS), which lets us precisely locate our position on the Earth given the distances to three out of twenty-four satellites orbiting our planet.

After measuring the distance from the epicenter to seismic station A, we know that the earthquake happened somewhere on the circle around A. But when we add the distance to seismic station B, we narrow the infinite number of options down to the two intersections of the circles around the two stations, like in Figure 8.6a. Finally, we measure the distance from the epicenter to station C, like in Figure 8.6b, which precisely locates the earthquake.

In practice, instead of intersecting at a single point, the three circles will create a spherical triangle. This is due to measurement errors and approximations, especially the assumption that the earthquake's focus and its epicenter are in the same place. In that case, we find the epicenter in the middle of the spherical triangle.

By the simple math of intersecting circles, to locate the earthquake, we need at least three seismic stations. But what if we have only one? This was the problem faced by seismologists analyzing seismic data from Mars.

*Without this assumption, in the rest of this section we'd replace circles with spheres, slightly complicating the math.

†If v_P and v_S are speeds of P and S waves, t_P and t_S are the times of their arrival after the earthquake, and d is the distance from the earthquake to the seismic station, then from the speed = distance / time formula we get $d = (t_S - t_P) / (1 / v_S - 1 / v_P)$.

‡Trilateration shouldn't be confused with triangulation, which, as we saw in Chapter 1, requires measuring angles.

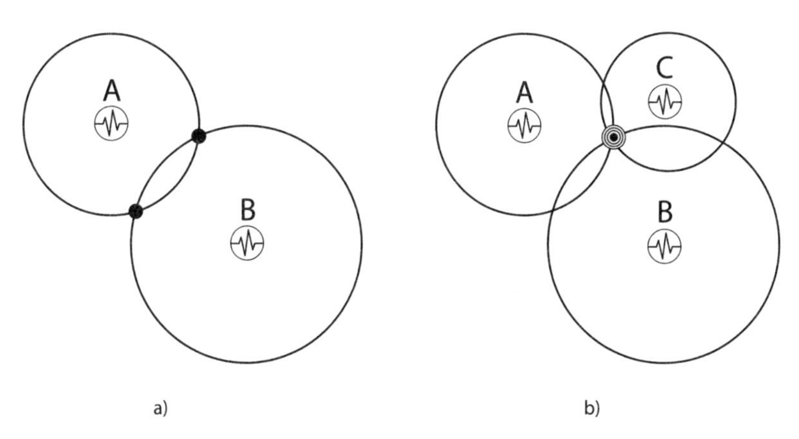

Figure 8.6 *a*) When we measure the distance to the epicenter from two observatories, we narrow down the earthquake's location to two points. *b*) Distances from the epicenter to three seismic stations are enough to find the earthquake's location.

InSight into Mars

Sending one seismometer to Mars was an ambitious project. Creating a network of seismometers, then, would be all but impossible. In 2018, NASA's *InSight* lander reached a flat, featureless, and quite boring part of the Red Planet's surface, aptly named Elysium Planitia, which is a combination of the ancient Greek *Elysium*—afterlife paradise—and the Latin *planitia*—flat surface. Equipped with a remarkably sensitive seismometer called Seismic Experiment for Interior Structure (SEIS), which was able to detect "surface movements smaller than a hydrogen atom," it had to take the place of a whole seismological network.

By the time *InSight* retired in December 2022, it had recorded 1,319 marsquakes. These quakes were so subtle that had they happened on the Earth, only people living close to the epicenter would have felt them, and damage would have been unlikely. Mars, unlike the Earth, doesn't have tectonic plates, so it doesn't show much seismic activity, and the quakes that occur there are likely due to meteorites crashing into the planet's surface or movements of magma. On the other hand, there's much less background noise than on the Earth, for example, because of the lack of oceans that on our planet are loud enough to be captured by seismometers thousands of miles away. Still, dealing with an instrument sensitive enough to detect the tiniest seismic activity made it hard to disentangle the signals from the noise. Cleaning up the signals required the input of many multidisciplinary teams of scientists, whose efforts paid off: two years into SEIS's operations, the first map of Mars's interior was published.

Just as on the Earth, the distance from SEIS to the quake's epicenter could be calculated from the difference between the arrival times of P waves and S waves—but, to pinpoint the epicenter, the scientists needed more information. Unlike on the Earth, they couldn't count on the data from other seismometers, so they came up with an alternative idea. SEIS measured seismic waves in all three dimensions, which allowed it to detect not only their occurrence and strength but also the direction they were coming from. These measurements weren't perfect, but the errors were relatively small, enabling the scientists to track the speeds and the paths of seismic waves. To replace the network of seismometers that on the Earth would capture many waves coming from the same earthquake, SEIS listened to the original waves and also to the echo—waves reflected off different layers of Mars. This complicated the math, but gave the scientists a much richer dataset, which—combined with insights from fields like geochemistry and mineralogy—enabled them to map Mars's interior.

In July 2021, the prestigious journal *Science* published three papers describing the new model of Mars's structure: one about the crust, one about the mantle, and one about the core. Before the mission, scientists had had some idea about Mars's interior, inferred from studies of Mars's gravitational field and meteorites that had fallen on Earth. Like Earth, Mars consists of a core surrounded by a mantle and a crust. With SEIS's readings, the researchers hoped to establish the thickness and density—and thus composition—of those layers. They compared thousands of possible maps—created, for example, by varying the thickness of the layers—to establish which ones best matched the observations. And they were in for some surprises, the largest of which concerned the core.

At a radius of about 1,830 kilometers—approximately half of Mars's total radius—the core is much larger than expected. Earlier, it had been assumed that Mars's core was made of similar materials as the Earth's core, mostly iron and nickel, which are denser than the rocky mantle. This would imply, however, that Mars, being a bit less dense than Earth, had a proportionally smaller core than our planet, which didn't match the data coming from SEIS. So, Mars's core must be made from different—lighter—materials than the Earth's. It's still unclear why. Solving this puzzle could help us understand the origin of rocky planets. We still don't know if Mars—like Earth—has an inner, liquid core, but measuring its size is already a huge achievement. Simon Stähler from ETH Zurich, the lead author of one of the three *Science* papers, commented: "It took scientists hundreds of years to measure Earth's core. After the Apollo missions, it took them forty years to measure the Moon's core. *InSight* took just two years to measure Mars' core." Who knows what we'll learn next?

Work in Progress

Finishing this chapter in a local cafe (yes, again), I can't help glancing at my phone. It knows not only which cafe I'm in, but which table I'm seated at. This impressive accuracy gives the illusion that maps of the Earth are already as good as they'll ever get. And yet, about half of our planet's surface remains unmapped—not to mention the great unknowns in the 260 billion cubic miles of the Earth's interior.

As opposed to most of the land surface, both the ocean floor and the Earth's interior remain invisible to our eyes and satellites. To reveal the secrets of those huge regions of our planet, we need to use waves other than light: sound waves to penetrate the oceans and seismic waves to "look" under the Earth's surface. While ancient cartographers had already made progress in mapping the continents, it was only a few decades ago that we developed the tools needed to map the hidden regions. These tools include not only technologies like sonar or advanced seismometers but also mathematical methods of converting numerical outputs into visual maps.

Mapping of the ocean floor and Earth's interior is still in progress; even as I'm writing this book, discoveries keep streaming in. Most recently, satellite measurements of gravitation revealed over 19,000 undersea volcanoes that remain to be mapped. A few weeks before that, two Australian seismologists found evidence for a distinct layer within the Earth's inner core. All these discoveries deepen our understanding of the Earth, satisfying our curiosity and helping us to protect both the planet and its inhabitants from the impacts of climate change, depletion of natural resources, and natural disasters. Mapping goes beyond filling the gaps in the atlases. It's hard to care for something we don't know, so mapping is an act of love for our planet—and ourselves.

CONCLUSION

How to Keep Up with Change

Sam rushes out of the door to get to the office in time. They pull out their phone to catch a ride, and in a few minutes, the car arrives. As they relax while the car slowly makes its way through San Francisco traffic, they don't even notice that something is missing: a driver.

Sam is one of my San Francisco-based friends who enthusiastically respond to my questions about their experiences with fully autonomous cars—that is, cars that can drive without a human in the front seat. They regularly ride with a Google-owned company, Waymo, because "they have enough money to hire really good people," which makes Sam feel safe.

In 2009, Google took on the challenge of developing autonomous cars and, after less than a decade, any interested resident or visitor in Phoenix, Arizona, could hail a self-driving car with a simple click on an app. Today, fully autonomous taxis are a popular mode of transportation in Phoenix and San Francisco due to their novelty, convenience, and low cost—although Sam points out that "you're paying for it with the risk that it could kill you." But they add that driverless cars seem to drive more safely than human drivers—if anything, the car will err on the side of caution, stopping completely when unsure how to proceed.

To replace a human driver's eyes, ears, and thinking patterns, a self-driving car combines information from multiple sources. Just as a person behind the wheel subconsciously merges her pre-existing knowledge of the area, her driving experience, and whatever she sees and hears in the moment, a machine uses very precise maps updated with real-time information from lidars,* radars, cameras, and other sensors.† Together, they allow the car to make quick decisions in any situation it might encounter on the road.

The prebuilt maps are the key, as they offload Waymo cars from having to scan for permanent features such as lane markers, road signs, or traffic lights, allowing them to focus on unexpected, tricky situations. In the same way, human drivers also feel much more comfortable driving in a familiar environment, where they can concentrate on watching out for pedestrians and surprising maneuvers of fellow drivers instead of making decisions at every turn and intersection. It's this overload of information that makes us much more tired after driving to a new holiday destination than after our daily commute.

Before a self-driving car is allowed in a new part of the city, the whole area must be mapped in unprecedented detail. GPS usually provides the location down to a few meters, which is enough for everyday use, but autonomous cars need to know their surroundings down to the centimeter so they can precisely locate details such as curbs, stop signs, and lane markers. Because of GPS's limitations, these detailed maps aren't created from satellite-based images but by lidar-equipped cars that drive through the area road by road, much as research boats map the ocean floor. The resulting three-dimensional images are later processed to create a map useful for the car, with features such as speed limits and lane markers often manually added by engineers. This kind of input is what self-driving cars prioritize to prepare for what lies ahead, which allows them, for example, to anticipate a stop sign or an intersection and start decelerating in advance, and thus drive smoothly. That's one of the many differences between human and machine drivers: people focus more on street names and characteristic buildings, without paying much attention to curbs or lanes.

As the car starts driving in the new area, it updates the inbuilt map with any fresh information it encounters. This becomes a bit of a chicken-and-egg problem: a car needs a map to locate itself within the area and it needs its location to alter the map accurately. Mathematicians have developed computational methods that allow for this seemingly impossible task

*Lidar (short for light detection and ranging technology) is based on the same principle as radar and sonar. It sends out light waves and measures the time it takes for the wave to return after encountering an obstacle.

†The technology varies between companies and develops quickly. I'm referring to Waymo cars at the time of writing.

of simultaneous localization and mapping (SLAM). This way, like a human driver, the car keeps improving its skills as it gathers new experiences on the road. The more situations it encounters, the better it's prepared to react when something unusual happens. Once again, a combination of maps and math is what keeps us safe.

As the world around us changes, maps change, too. An increased interest in sailing (or, rather, greed for overseas territories) inspired Mercator's flawed but revolutionary map suited for navigation. The rapidly growing transportation network prompted Henry Beck to create a geographically inaccurate but exceptionally useful topological map of the London Underground. And the advent of self-driving cars brings about technological advances that allow the production of extremely precise local maps.

Our past, present, and future determine the maps we need. The maps we make, on the other hand, have changed history—and will inevitably change our future. A map of an epidemic can save lives; a map of school districts can facilitate the future success of some children and hinder others; a map of the ocean floor can help us ameliorate the effects of climate change.

Behind every map, there's mathematics. It's mathematics that helps cartographers to choose the most appropriate projection by explaining the distortions involved. It's mathematics that allows delivery companies to plan the optimal route for their drivers. It's mathematics that assists us when we navigate mazelike cities. Whenever we open a map, we hold in our hands the results of centuries of mathematical research. Self-driving cars would probably remain in the realm of science fiction without Gauss's research on curvatures, Snell's advances in triangulation, and Euler's invention of graphs—all of which happened centuries ago. Vast areas of our planet would still be an enigma without the extraordinary work of two exceptional women—Inge Lehmann and Marie Tharp—and many others who followed in their footsteps.

Mathematicians and mapmakers have influenced not only how we see the world but also how we function within it. With smartphones in our pockets, we take maps for granted, forgetting all the extraordinary science and technology that has brought them into existence. All maps are flawed, and yet our reality would be completely different without them. Necessarily imperfect, maps make the life we're living possible.

NOTES

FURTHER READING

ACKNOWLEDGMENTS

ILLUSTRATION CREDITS

INDEX

NOTES

Introduction

Page 2 it is widely known to be deceptive to the eye
John Joseph O'Connor and Edmund Frederick Robertson, "Gerardus Mercator," MacTutor (Maths History), School of Mathematics and Statistics, University of St Andrews, August 2002, https://mathshistory.st-andrews.ac.uk/Biographies/Mercator_Gerardus/.

2 we keep an online version of it in our pockets
Sarah E. Battersby et al., "Implications of Web Mercator and Its Use in Online Mapping," *Cartographica* 49, no. 2 (2014): 85–101, https://doi.org/10.3138/carto.49.2.2313.

1. Curved

5 1828 fictional biography History of the Life and Voyages of Christopher Columbus
V. Frederick Rickey, "How Columbus Encountered America," *Mathematics Magazine* 65, no. 4 (1992): 219–225, https://doi.org/10.1080/0025570x.1992.11996024.

5 a spherical Earth's shadow on the Moon
Jill Howard, "Shaping the Universe I: Planet Earth," NRICH, University of Cambridge, 2007, rev. 2011, https://nrich.maths.org/5732.

6 the Earth must be small
Rickey, "How Columbus Encountered America."

6 a small sea navigable in a few days
Samuel Eliot Morison, *Journals and Other Documents on the Life and Voyages of Christopher Columbus* (New York: Heritage Press, 1963), 22.

8 considered the founder of scientific geodesy
I. K. Fischer, "Geodesy: Historical Introduction," in *Geophysics: Encyclopedia of Earth Science* (Boston: Springer, 1989).

8 the Earth's shape, orientation in space, and gravitational field
"What Is Geodesy?" National Ocean Service, NOAA, 1 March 2014, updated 6 November 2023, https://oceanservice.noaa.gov/facts/geodesy.html#:~:text=Geodesy%20is%20the%20science%20of%20accurately%20measuring%20and%20understanding%20three,of%20these%20properties%20with%20time.

9 ancient and medieval geographers
Samuel Eliot Morison, *Admiral of the Ocean Sea: A Life of Christopher Columbus* (Boston: Little, Brown, 1942).

9 the fastest route to the "land of spices," that is, India
Helen Wallis, "What Columbus Knew," *History Today,* 42, no. 5 (5 May 1992): 17–23, https://www.historytoday.com/archive/what-columbus-knew.

9 going westward was a better idea
Mary Ames Mitchell, "Marco Polo and Polo [sic] Toscanelli," Crossing the Ocean Sea—Marco Polo and Paulo Toscanelli, 2015, https://web.archive.org/web/20230309223459/https://crossingtheoceansea.com/OceanSeaPages/OS-49-PoloToscanelli.html.

10 is only about 200 kilometers
"Japan: A Geographical Sketch," Asia Society, 2002, http://sites.asiasociety.org/education/AsianArt/other.geo.japan.htm#:~:text=The Japanese archipelago lies off,peninsula as the nearest landmass.

10 Columbus reduced the Earth's circumference by about a quarter
Morison, *Admiral of the Ocean Sea,* 54–78.

10 the Virgin Islands in the Caribbean
Morison, *Admiral of the Ocean Sea,* 54–78.

10 the enthusiasm of the Spanish treasurer
Valerie I. J. Flint, "Christopher Columbus," *Encyclopaedia Britannica,* 26 July 1999, last updated 19 October 2023, https://www.britannica.com/biography/Christopher-Columbus.

11 the westward route to the Indies
"Christopher Columbus," Royal Museums Greenwich, n.d., accessed 17 May 2023, https://www.rmg.co.uk/stories/topics/christopher-columbus.

11 he had expected to reach Cipangu
Rickey, "How Columbus Encountered America."

11 it wasn't a huge success
John Joseph O'Connor and Edmund Frederick Robertson, "Petrus Apianus," MacTutor (Maths History), School of Mathematics and Statistics, University of St Andrews, April 2002, https://mathshistory.st-andrews.ac.uk/Biographies/Apianus/.

13 the idea of triangulation quickly spread through Europe
Shaw Kinsley et al., "*Cosmographia:* A Close Encounter," Student Virtual Exhibition, Museum of Science, Oxford, 1999. See section "The Book: Many Versions of *Cosmographia,*" accessed 17 May 2023, https://www.mhs.ox.ac.uk/students/98to99/Book/Bookpgs/PageTwo.html.

13 We'll meet him again in the next chapter
Thony Christie, "Mapping the History of Triangulation," The Renaissance Mathematicus blogpost, 25 May 2012, https://thonyc.wordpress.com/2012/05/25/mapping-the-history-of-triangulation/.

13 which translates to The Dutch Eratosthenes
Fokko Jan Dijksterhuis, "The Mutual Making of Sciences and Humanities: Willebrord Snellius, Jacob Golius, and the Early Modern Entanglement of Mathematics and Philology," in *The Making of the Humanities,* vol. 2, ed. Rens Bod et al. (Amsterdam: University of Amsterdam Press, 2012), 73–92, https://doi.org/10.1017/9789048517336.005.

13 about eighty miles almost directly south
John Joseph O'Connor and Edmund Frederick Robertson, "Willebrord van Royen Snell," MacTutor (Maths History), School of Mathematics and Statistics, University of St Andrews, November 2010, https://mathshistory.st-andrews.ac.uk/Biographies/Snell/.

13 within 4 percent of the modern estimate
Thony Christie, "Getting the Measure of the Earth," The Renaissance Mathematicus blogpost, 22 July 2013, https://thonyc.wordpress.com/2013/07/22/getting-the-measure-of-the-earth/.

13 mathematical tools he had at his disposal
Matthew H. Edney, "Geodetic Surveying in the Enlightenment," in *History and Cartography,* vol. 4, *Cartography in the European Enlightenment,* ed. Matthew H. Edney and Mary S. Pedley (Chicago: University of Chicago Press, 2019), 439–450.

14 Measure of the Earth
Larrie D. Ferreiro, *Measure of the Earth: The Enlightenment Expedition That Reshaped Our World* (New York: Basic Books, 2013).

16 Spain insisted that "two intelligent Spaniards accompany the said scientists"
Ferreiro, *Measure of the Earth,* 38.

17 within 120 yards of the currently accepted value
Ferreiro, *Measure of the Earth,* 328.

18 an inspiration for his liberation movement
Douglas Main, "The Remarkable Story of the First Accurate Measure of the Earth," *Popular Mechanics,* 3 June 2011, https://www.popularmechanics.com/science/environment/a6674/the-remarkable-story-of-the-first-accurate-measure-of-the-earth/.

18 Ecuador means simply "Republic of the Equator"
Main, "Remarkable Story."

18 from paper maps to today's GPS
Ferreiro, *Measure of the Earth,* 287.

18 poor parents without much formal education
G. Waldo Dunnington and Jeremy Gray, *Gauss: Titan of Science* (Washington, DC: Mathematical Association of America, 2004), 8.

18 his exceptional talents
Walter K. Bühler, *Gauss: A Biographical Study* (New York: Springer, 1981).

18 add numbers from one to one hundred
Brian Hayes, "Gauss's Day of Reckoning: A Famous Story about the Boy Wonder of Mathematics Has Taken on a Life of Its Own," *American Scientist* 94, no. 3 (2006): 200, https://www.americanscientist.org/article/gausss-day-of-reckoning.

19 the future "Prince of Mathematics"
Dunnington and Gray, *Gauss,* 337.

19 applying mathematics to solve real-world problems
W. Freeden and M. Zuhair Nashed, *Handbook of Mathematical Geodesy Functional Analytic and Potential Theoretic Methods* (Cham, Switzerland: Springer International, 2018).

19 "the most refined geometer and the perfect astronomer"
Dunnington and Gray, *Gauss,* 113.

19 his early interest in geodesy
Fred Roeder, "Carl Friedrich Gauss," *Backsights,* Surveyors Historical Society, 1993, http://www.surveyhistory.org/carl_friedric.htm.

19 ten-mark banknote celebrating his life and work
Oleg Leyderman, "10 Mark 1993, Germany," Notes Collection, n.d., accessed 17 May 2023, https://notescollector.eu/pages/en/notes.php?noteId=1096.

23 making it harder for the wind to change its shape
Zhi Sun et al., "The Mechanical Principles behind the Golden Ratio Distribution of Veins in Plant Leaves," *Scientific Reports* 8, no. 1 (September 14, 2018), https://doi.org/10.1038/s41598-018-31763-1.

23 to keep the folded shape
Przemyslaw Prusinkiewicz and Pierre Barbier de Reuille, "Constraints of Space in Plant Development," *Journal of Experimental Botany* 61, no. 8 (May 13, 2010): 2117–2129, https://doi.org/10.1093/jxb/erq081.

24 the strength of the curvy egg shape
Eric N. Hahn et al., "Nature's Technical Ceramic: The Avian Eggshell," *Journal of The Royal Society Interface* 14, no. 126 (January 1, 2017): 20160804, https://doi.org/10.1098/rsif.2016.0804.

25 increasing the sensation of crunchiness
Kathleen Villaluz, "Crunchy Engineering of Pringles' Hyperbolic Paraboloid Shape," Interesting Engineering website, 24 December 2020, https://interestingengineering.com/culture/geometry-of-pringles-crunchy-hyperbolic-paraboloid.

25 she spots a bear standing next to her tent
Boris Khesin and Serge Tabachnikov, "Polar Bear or Penguin? Musings on Earth Cartography and Chebyshev Nets," *Mathematical Intelligencer* 43, no. 1 (2020): 20–24, https://doi.org/10.1007/s00283-020-10013-1.

29 maybe unexpectedly, the Cold War
"Why Anchorage Is America's Most OP City," RealLifeLore, YouTube video, 24:33, 22 December 2021, https://www.youtube.com/watch?v=UMNfagIz0hs.

29 between Singapore and New York
Ethan Klapper and Zach Griff, "The World's Longest Nonstop Flights, Updated," The Points Guy, accessed 14 June 2023, https://thepointsguy.com/news/the-worlds-longest-nonstop-flights-updated/.

30 the connecting flight to their destination
Kevin Bonsor, "How Airlines Work: Hubs and Spokes," HowStuffWorks, 2001, https://science.howstuffworks.com/transport/flight/modern/airline3.htm.

30 the world's busiest international airport is in Dubai
"Busiest Airports in the World," OAG, accessed 17 May 2023, https://www.oag.com/busiest-airports-world.

30 the relatively small city of Helsinki, the capital of Finland
Becca Rowland, "Great Circle Routes & Flight Paths," OAG, 6 November 2019, https://www.oag.com/blog/great-circle-routes-flight-paths.

30 his rapping nephew's song "Flat to Fact"
Laura Wagner, "Neil DeGrasse Tyson Gets into a Rap Battle with B.o.B over Flat Earth Theory," NPR, 26 January 2016, https://www.npr.org/sections/thetwo-way/2016/01/26/464474518/neil-degrasse-tyson-gets-into-a-rap-battle-with-b-o-b-over-flat-earth-theory.

2. Flat

32 on which everything looked different
Bob Shaffer, "Arriving at Boston Public Schools: More Accurate—and Inclusive—World Maps," WBUR News, 20 March 2017, https://www.wbur.org/news/2017/03/16/world-maps-boston-public-schools.

34 unwrap the paper to get your map projection
John P. Snyder, *Map Projections: A Working Manual,* U.S. Geological Survey Professional Paper 1395 (Washington, DC: US Government Printing Office, 1987), https://pubs.usgs.gov/pp/1395/report.pdf.

36 How to Write the Latin Letters Which They Call Italic or Cursive
Mark S. Monmonier, *Rhumb Lines and Map Wars: A Social History of the Mercator Projection* (Chicago: University of Chicago Press, 2004), 32.

36 engraving labels on their collaborative creations

John J. O'Connor and Edmund F. Robertson, "Gerardus Mercator," MacTutor (Maths History), School of Mathematics and Statistics, University of St Andrews, August 2002, https://mathshistory.st-andrews.ac.uk/Biographies/Mercator_Gerardus/.

37 before anyone else (as in, any European) got in there first

Nuno Crato, *Figuring It Out: Entertaining Encounters with Everyday Math* (Heidelberg: Springer Science & Business Media, 2010), 66.

39 In the map's verbose legend

"G. Mercator, B. Van Hoff, and C. Nooteboom, Gerard Mercator's Map of the World (1569) in the Form of an Atlas in the Maritiem Museum 'Prins Hendrik' at Rotterdam; Reproduced on the Scale of the Origional and Issued by the Maritiem Museum . . . and the Editors of Imago Mundi; with an Introduction by Van Hoff; Foreword by C. Nooteboom (Maritiem Museum, 1961)."

39 the lengthening of the parallels with reference to the equator

V. Frederick Rickey and Philip M. Tuchinsky, "An Application of Geography to Mathematics: History of the Integral of the Secant," *Mathematics Magazine* 53, no. 3 (1980): 162–166, https://doi.org/10.1080/0025570x.1980.11976846.

40 A new and enlarged description of the Earth with corrections for use in navigation

Robert Osserman, "Mathematical Mapping from Mercator to the Millennium," in *Mathematical Adventures for Students and Amateurs,* ed. David F. Hayes and Tatiana Shubin (Providence, RI: Mathematical Association of America, 2004).

40 202 × 124 cm (80 × 49 in.)

Monmonier, *Rhumb Lines and Map Wars,* 4.

40 a surprising mathematical result

Rickey and Tuchinsky, "Application of Geography to Mathematics."

40 In Certaine Errors in Navigation

Edward Wright, *Certaine Errors in Navigation, Arising Either of the Ordinarie Erroneous Making or Using of the Sea Chart, Compasse, Crosse Staffe, and Tables of Declination of the Sunne, and Fixed Starres Detected and Corrected* (London: Sims, 1599).

41 "An Application of Geography to Mathematics: History of the Integral of the Secant"

Rickey and Tuchinsky, "Application of Geography to Mathematics."

43 "teacher of navigation, survey and other parts of the mathematics"

Rickey and Tuchinsky, "Application of Geography to Mathematics," 164.

46 regions that are far from the equator

Andrew Taylor, *The World of Gerard Mercator: The Mapmaker Who Revolutionised Geography* (London: Harper Perennial, 2005).

46 *Alaska is smaller than Mercator's inconspicuous Libya*
Betsy Mason, "Why Your Mental Map of the World Is Wrong," *National Geographic,* 3 May 2021, https://www.nationalgeographic.com/culture/article/all-over-the-map-mental-mapping-misconceptions.

46 *a large part of Europe within its boundaries*
Mark Fischetti, "Africa Dwarfs China, Europe and the U.S.," *Scientific American,* 1 July 2015, https://www.scientificamerican.com/article/africa-dwarfs-china-europe-and-the-u-s/.

46 *to both Queen Mary and Queen Elizabeth I*
"John Dee's Petition to James I Asking to Be Cleared of Accusations of Conjuring, 1604," British Library, accessed 8 November 2021, https://www.bl.uk/collection-items/john-dees-petition-to-james-i-asking-to-be-cleared-of-accusations-of-conjuring-1604.

46 *allegedly even came up with the term "British Empire"*
William H. Sherman, *John Dee: The Politics of Reading and Writing in the English Renaissance* (Amherst, MA: University of Massachusetts Press, 1995).

47 *working for a London bookseller*
Taylor, *World of Mercator,* 198.

47 *one of the youngest filmmakers in Germany*
Bob Abramms, "Arno Peters: Photo Biography," Many Ways to See the World, ODT Maps, accessed 12 November 2020, https://web.archive.org/web/20201112004208/https://manywaystoseetheworld.org/pages/arno-peters-photo-biography.

47 *shifting from film production to history and social studies*
Monmonier, *Rhumb Lines and Map Wars,* 147.

48 *to better understand world history*
Abramms, "Arno Peters: Photo Biography."

48 *starting with its authorship*
Monmonier, *Rhumb Lines and Map Wars,* 148.

48 *different equal-area projections*
Snyder, *Map Projections,* 48.

49 *"ragged, long, winter underwear hung out to dry on the Arctic circle"*
Arthur H. Robinson, "Arno Peters and His New Cartography," *American Cartographer* 12, no. 2 (1985): 103–111, https://doi.org/10.1559/152304085783915063.

49 *only one in two students*
"Boston Public Schools at a Glance 2019–2020," Boston Public Schools Communications Office, December 2019, https://www.bostonpublicschools.org/cms/lib/MA01906464/Centricity/Domain/187/BPS%20at%20a%20Glance%202019-20_FINAL.pdf.

49 "it was amazingly interesting to see [the students] questioning what they thought they knew"

Joanna Walters, "Boston Public Schools Map Switch Aims to Amend 500 Years of Distortion," *The Guardian,* 23 March 2017.

50 its most accurate representation: the globe

Josh Hostetter (@bosstetter_edu), "Mercator Projection? Gall-Peters Projection? Nah. They're both distorted. What about . . . #BuyRealGlobes; http://businessinsider.com/boston-school-gall-peters-map-also-wrong-mercator-2017-3 via @sai," Twitter, 1 August 2017, 9:21 p.m., https://twitter.com/bosstetter_edu/status/892465287836180481?s=20.

50 impacted students' worldviews

Thomas Saarinen, Michael Parton, and Roy Billberg, "Relative Size of Continents on World Sketch Maps," *Cartographica* 33, no. 2 (1996): 37–48, https://doi.org/10.3138/f981-783n-123m-446r.

50 from twenty-two cities

The cities were: Abidjan (Ivory Coast), Antananarivo (Madagascar), Rabat (Morocco), Ruhengeri (Rwanda), Stellenbosch (South Africa), Hong Kong, Kuwait, Pune (India), Seoul (South Korea), Silpakorn (Thailand), Armidale (Australia), Istanbul (Turkey), Lisbon (Portugal), Trondheim (Norway), Fairbanks (United States), Nassau (Bahamas), Ottawa (Canada), Seattle (United States), Tucson (United States), Belo Horizonte (Brazil), Buenos Aires (Argentina), and Caracas (Venezuela).

50 Ghent University didn't find the "Mercator effect"

Lieselot Lapon, Kristien Ooms, and Philippe De Maeyer, "The Influence of Map Projections on People's Global-Scale Cognitive Map: A Worldwide Study," *ISPRS International Journal of Geo-Information* 9, no. 4 (2020): 196, https://doi.org/10.3390/ijgi9040196.

52 which made the directions less precise

Rangi42, "Five world map styles," Hacker News, 29 March 2017, https://web.archive.org/web/20240203163845/https://news.ycombinator.com/item?id=13986412.

53 a public registry of all things geographical

"EPSG Geodetic Parameter Dataset," GeoRepository, accessed 23 May 2023, https://epsg.org/home.html.

53 which is "Google" transliterated to numbers

Christopher Schmidt, "Google Projection: 900913," *Technical Ramblings: Ramblings of a Hacker,* 7 August 2007, https://crschmidt.net/blog/archives/243/google-projection-900913/.

53 the confusingly similar EPSG:3857

Alastair Aitchison, "The Google Maps / Bing Maps Spherical Mercator Projection," 23 January 2011, https://alastaira.wordpress.com/2011/01/23/the-google-maps-bing-maps-spherical-mercator-projection/.

53 "more user-friendly, especially for those dealing with multiple data sources"
Sarah E. Battersby et al., "Implications of Web Mercator and Its Use in Online Mapping," *Cartographica* 49, no. 2 (2014): 85–101, https://doi.org/10.3138/carto.49.2.2313.

53 a big change to their maps
Google Maps (@googlemaps), "With 3D Globe Mode on Google Maps desktop, Greenland's projection is no longer the size of Africa. Just zoom all the way out at http://google.com/maps," Twitter, 2 August 2018, 11:26 p.m., https://twitter.com/googlemaps/status/1025130620471656449?s=20.

54 the oceans uninterrupted
John W. Hessler, "Robinson Projection," in *The History of Cartography,* vol. 6, *Cartography in the Twentieth Century,* ed. Mark Monmonier (Chicago: University of Chicago Press, 2015), 1357–1358.

54 there are hundreds of named projections
Tobias Jung, "My Projection Collection: Available Map Projections," Compare Map Projections, last update 16 June 2023, accessed 23 December 2021, https://map-projections.net/projections-list.php.

55 describe his projection mathematically
Hessler, "Robinson Projection."

55 where it stayed for over a decade
Monmonier, *Rhumb Lines and Map Wars,* 134.

55 newspapers followed the magazine's lead
Monmonier, *Rhumb Lines and Map Wars,* 136.

56 another cartographer
Barbara Bartz Petchenik: "Women Who Shaped the World: Barbara Petchenik," The Future Mapping Company, 26 September 2019, https://futuremaps.com/blogs/news/women-cartographers-barbara-petchenik.

56 the philosophy of map understanding
Arthur Howard Robinson and Barbara Bartz Petchenik, *The Nature of Maps: Essays toward Understanding Maps and Mapping* (Chicago: University of Chicago Press, 1976).

56 I'm sure this would have made her proud
Will C. van den Hoonaard, *Map Worlds: A History of Women in Cartography* (Waterloo, Ontario, Canada: Wilfrid Laurier University Press, 2013).

56 "It points so far north," he commented in disbelief
Reprinted from *Washington Daily News* in Don May, "You Can't Build That Mosque with a Compass," *Surveying and Mapping* 13, no. 1 (January 1953): 367–368.

57 which isn't the same as the "true," geographic north
Erika Jones, "True North and Magnetic North: What's the Difference?" Royal Museums Greenwich, n.d., accessed 11 July 2023, https://www.rmg.co.uk/stories/topics/true-north-magnetic-north-whats-difference.

58 is generally considered the accepted qibla
Daniel Z. Levina, "Which Way Is Jerusalem? Which Way Is Mecca? The Direction-Facing Problem in Religion and Geography," *Journal of Geography* 101, no. 1 (2002): 27–37, https://doi.org/10.1080/00221340208978464.

58 today known as the Craig retroazimuthal or Mecca projection
Waldo Tobler, "Qibla, and Related, Map Projections," *Cartography and Geographic Information Science* 29, no. 1 (2002): 17–23, https://doi.org/10.1559/152304002782064574.

58 the mecca of the financial world
Tobler, "Qibla, and Related Map Projections."

60 an article on the North Korean threat
"When Bluff Turns Deadly," *The Economist,* 1 May 2003.

60 Something didn't add up
Thierry Gregorius, "Distances on a World Map: The Classic Geodetic Blunder," Georeferenced blog, 22 May 2014, https://georeferenced.wordpress.com/2014/05/22/worldmapblunders/.

61 thus destroying our peace of mind
"Correction: North Korea's Missiles', *The Economist,* 15 May 2003.

61 the danger of a missile attack on the United States
"How Potent Are North Korea's Threats?" BBC News, 15 September 2015, https://www.bbc.com/news/world-asia-21710644.

61 it was certainly misleading
Jason Davies, "Distances from North Korea," Maps, n.d., accessed 30 August 2021, https://www.jasondavies.com/maps/north-korea-distance/.

61 So many messages in one simple map!
Arthur Jay Klinghoffer, *The Power of Projections: How Maps Reflect Global Politics and History* (Westport, CT: Praeger, 2006), 111.

62 just off Sakhalin Island in today's Russia
Thom Patterson, "KAL Flight 007: A Cold War-Fueled Tragedy," CNN, 31 August 2013, https://edition.cnn.com/2013/08/31/us/kal-fight-007-anniversary/index.html.

62 the Soviets' reaction was considered unreasonable
Xinyue Wu, "Map of Korean Airlines Flight 007," Mappenstance, 3 April 2018, https://blog.richmond.edu/livesofmaps/2018/04/03/map-of-korean-airlines-flight-007/.

62 despite the tragedy of lives lost
Patricia Gilmartin, "The Design of Journalistic Maps: Purposes, Parameters and Prospects," *Cartographica* 22, no. 4 (1985): 1–18, https://doi.org/10.3138/4n0w-35p7-w416-9341.

3. Scaled

65 mathematician and photographer Lewis Carroll
John J. O'Connor and Edmund F. Robertson, "Charles Lutwidge Dodgson," MacTutor (Maths History), School of Mathematics and Statistics, University of St Andrews, November 2002, https://mathshistory.st-andrews.ac.uk/Biographies/Dodgson/.

65 according to Mein Herr, worried the farmers
Lewis Carroll, "The Man in the Moon," in *Sylvie and Bruno Concluded* (New York: Macmillan, 1894).

66 equivalent to the nominal scale
Miljenko Lapaine and E. Lynn Usery, *Choosing a Map Projection* (Cham, Switzerland: Springer, 2017).

67 the Baltimore phenomenon
Michael P. Peterson, *Mapping in the Cloud* (New York: Guilford Press, 2014).

67 almost a quarter of all international borders follow rivers
Adam Voiland, "How Rivers Shape States," NASA Earth Observatory, 8 September 2020, https://earthobservatory.nasa.gov/images/147242/how-rivers-shape-states.

67 didn't stop him from enslaving over six hundred people
"Slavery at Monticello: Slavery FAQs: Property," Monticello, Thomas Jefferson Foundation, n.d., accessed 9 September 2021, https://www.monticello.org/slavery/slavery-faqs/property.

68 the Rocky Mountain states (Colorado, Montana, and Wyoming)
Erin Allen, "States and Their Shapes," *Library of Congress Information Bulletin* 67, no. 9 (September 2008), https://www.loc.gov/loc/lcib/0809/index.html.

68 it turns out that Colorado has 697 sides
Frank Jacobs, "Colorado Is Not a Rectangle—It Has 697 Sides," Big Think, 31 October 2018, updated January 2023, https://bigthink.com/the-present/colorado-is-not-a-rectangle/#rebelltitem6.

69 describe interactions between countries
John J. O'Connor and Edmund F. Robertson, "Lewis Fry Richardson," MacTutor (Maths History), School of Mathematics and Statistics, University of St Andrews, October 2003, https://mathshistory.st-andrews.ac.uk/Biographies/Richardson/.

69 in other words, on deadly quarrels
Lewis Fry Richardson, *Statistics of Deadly Quarrels,* ed. Quincy Wright and C. C. Lienau (Pittsburgh, PA: Boxwood Press, 1960).

69 Spain estimated the same border as 987 kilometers
P. G. Drazin, "Fractals," in *The Collected Papers of Lewis Fry Richardson,* vol. 1, ed. Oliver M. Ashford et al. (Cambridge: Cambridge University Press, 1993), 45–46.

71 the oldest professor to receive tenure at Yale
Steve Olson, "The Genius of the Unpredictable," *Yale Alumni Magazine,* November / December 2004, http://archives.yalealumnimagazine.com/issues/2004_11/mandelbrot.html.

71 took care of his education
John J. O'Connor and Edmund F. Robertson, "Benoit Mandelbrot," MacTutor (Maths History), School of Mathematics and Statistics, University of St Andrews, July 1999, https://mathshistory.st-andrews.ac.uk/Biographies/Mandelbrot/.

71 the École Polytechnique
Benoît Mandelbrot, "École Normale and Thought in Mathematics," Web of Stories video, 24 January 2008, https://www.webofstories.com/play/benoit.mandelbrot/16.

71 (and getting married in the meantime)
Nigel Lesmoir-Gordon, "Benoît Mandelbrot Obituary," *The Guardian,* 17 October 2010.

72 very complicated and very simple at the same time
Benoit Mandelbrot, "Fractals and the Art of Roughness," TED Talks video, February 2010, https://www.ted.com/talks/benoit_mandelbrot_fractals_and_the_art_of_roughness?language=en.

72 no possible significance
Benoît Mandelbrot, "Errors of Transmission in Telephone Channels (50 / 144)," Web of Stories video, n.d., https://www.youtube.com/watch?v=0EeAclc1OEc.

72 as IBM's engineers tended to explain the problem
James Gleick, "A Geometry of Nature," in *Chaos: Making a New Science* (Harrisonburg, VA: Viking, 1987), 81–118.

73 the General Systems Year Book *of 1961*
Oliver M. Ashford, *Prophet or Professor: The Life and Work of Louis Fry Richardson* (Bristol: Adam Hilger, 1985), 260, quoted in J. C. R. Hunt, "Lewis Fry Richardson and His Contributions to Mathematics, Meteorology, and Models of Conflict," *Annual Review of Fluid Mechanics* 30, no. 1 (January 1998): xiii–xxxvi, https://doi.org/10.1146/annurev.fluid.30.1.0.

73 incomprehensible modern monographs?
"Benoit Mandelbrot," interview by Anthony Barcellos, in *Mathematical People: Profiles and Interviews,* ed. Donald J. Albers and Gerald L. Alexanderson, 2nd ed. (Wellesley, MA: A. K. Peters, 2008), 213–232.

78 covered by a grid of tiny squares
Sven Kreiss, "S2 Cells and Space-Filling Curves: Keys to Building Better Digital Map Tools for Cities," Sidewalk Talk, 29 July 2016, https://medium.com/sidewalk-talk/s2-cells-and-space-filling-curves-keys-to-building-better-digital-map-tools-for-cities-a312aa5e2f59.

78 looks over Rio de Janeiro from 41204668869
S2 Region Coverer Online Viewer app, accessed 4 February 2024, https://igorgatis.github.io/ws2/?cells=00997fd59c5.

79 purposefully incomprehensible
Benoit Mandelbrot, "How Long Is the Coast of Britain? Statistical Self-Similarity and Fractional Dimension," *Science* 156, no. 3775 (May 5, 1967): 636–638, https://doi.org/10.1126/science.156.3775.636.

80 Mandelbrot brought fractal dimensions to a wider audience
Mandelbrot, "How Long Is the Coast of Britain?"

80 the length read off the school map
Hugo Steinhaus, "Length, Shape and Area," *Colloquium Mathematicum* 3, no. 1 (1954): 1–13, https://doi.org/10.4064/cm-3-1-1-13.

80 Sicilian fields were known to have bigger areas
Benoît Mandelbrot, "A Geometry Able to Include Mountains and Clouds," in *The Colours of Infinity: The Beauty, the Power of Fractals,* ed. Nigel Lesmoir-Gordon (London: Springer, 2010), 38–57.

80 leave me personally offended
Mandelbrot, "How Long Is the Coast of Britain?"

81 a mountain behind it
"Hunting the Hidden Dimension," NOVA Transcripts, Public Broadcasting Service, 28 October 2008, https://www.pbs.org/wgbh/nova/transcripts/3514_fractals.html.

82 the results stunned even him
Rachel Sullivan, "Hiding in Plain Sight," *Discovery Channel Magazine,* 2004, https://cpb-us-e1.wpmucdn.com/blogs.uoregon.edu/dist/e/12535/files/2017/02/Hiding-in-Plain-Sight-z3v4wy.pdf.

82 Alvy Ray Smith commented on his decision
Tekla S. Perry, "And the Oscar Goes to . . . ," IEEE Spectrum, 2 April 2001, https://spectrum.ieee.org/and-the-oscar-goes-to.

83 free noncommercial use
"What Is Renderman?" Pixar, n.d., accessed 7 September 2021, https://renderman.pixar.com/about.

83 nor does lightning travel in a straight line
Benoit B. Mandelbrot, "Introduction: Theme," in *The Fractal Geometry of Nature* (San Francisco: W. H. Freeman, 1983; New York: W. H. Freeman, 2021), 1–5.

83 the United States and Russian territories
"How Close Is Alaska to Russia?," Alaska Public Lands, National Park Service, accessed 4 February 2024, https://www.nps.gov/anch/learn/historyculture/how-close-is-alaska-to-russia.htm.

83 the native Łingít and Haida communities
D. M. L. Farr and Niko Block, "Alaska Boundary Dispute," The Canadian Encyclopedia, 6 February 2006, last edited 9 August 2016, https://thecanadianencyclopedia.ca/en/article/alaska-boundary-dispute.

85 forty-six by the Americans
"Mapping the Alaska Boundary Dispute: A Cartographic History of the British-United States Border Tribunal of 1903," Arctic and Northern Studies StoryMap, University of Alaska Fairbanks, n.d., accessed 27 December 2023, Story Map Cascade, Environmental Systems Research Institute (ESRI), Redlands, CA, https://www.arcgis.com/apps/Cascade/index.html?appid=9ab6efdafeb841d480d1933dac12a622.

85 forced to cede the harbor they had fought for
"Mapping the Alaska Boundary Dispute."

85 which mountains the treaty was talking about
"The Alaska Boundary Case (Great Britain, United States)," United Nations, Reports of International Arbitral Awards, 20 October 1903, vol. 15, 481–540, https://legal.un.org/riaa/cases/vol_XV/481-540.pdf.

86 more autonomy from the Crown
Farr and Block, "Alaska Boundary Dispute."

86 coastal nations should own the most adjacent waters
Andrew Blom, "Hugo Grotius (1583–1645)," Internet Encyclopedia of Philosophy, n.d., accessed 23 November 2021, https://iep.utm.edu/grotius/.

86 who these resources belong to
Ryan B. Stoa, "The Coastline Paradox," *Rutgers University Law Review* 72, no. 2 (2019): 351–400, https://rutgerslawreview.com/wp-content/uploads/2020/11/72_Rutgers_Univ_L_Rev_0351_Stoa.pdf.

87 straight lines when it's impractical
"The United Nations Convention on the Law of the Sea (A Historical Perspective)," Oceans & Law of the Sea, United Nations, originally prepared 1998, accessed 21 August 2021, https://www.un.org/depts/los/convention_agreements/convention_historical_perspective.htm.

88 which started a fishing war between the two countries
Anglo-Norwegian Fisheries, UK v. Norway, Order, 1951 I.C.J. 117 (Jan. 18).

89 the larger your tax
Stoa, "Coastline Paradox."

4. Distanced

90 come up with alternative routes
Shaun Larcom, Ferdinand Rauch, and Tim Willems, "The Benefits of Forced Experimentation: Striking Evidence from the London Underground Network,"

Quarterly Journal of Economics 132, no. 4 (2017): 2019–2055, https://doi.org/10.1093/qje/qjx020.

91 it required a separate map
"Mapping London: The Iconic Tube Map," London Transport Museum, n.d., accessed 29 May 2023, https://www.ltmuseum.co.uk/collections/stories/design/mapping-london-iconic-tube-map.

92 Beck sketched a new version
Max Roberts, "Tube Map Myths: The Diagrammatic London Underground Map Was Inspired by Electrical Circuit Diagrams," Tube Map Central, 30 November 2007, http://tubemapcentral.com/writing/webarticles/myths/mythset/02circuit.html.

92 Connections are the thing
"The London Underground Map Narrated by Jancis Robinson," Design Classics 5, BBC Worldwide video, 1987, https://search.alexanderstreet.com/preview/work/bibliographic_entity%7Cvideo_work%7C1796780.

92 printed on 500 trial pocket maps
"Harry Beck," Famous Graphic Designers, 2018, accessed 29 May 2023, https://www.famousgraphicdesigners.org/harry-beck.

92 clear and simple to navigate
Alexander J. Kent, "When Topology Trumped Topography: Celebrating 90 Years of Beck's Underground Map," *Cartographic Journal* 58, no. 1 (2021): 1–12, https://doi.org/10.1080/00087041.2021.1953765.

92 which didn't seem so distant anymore
Darien Graham-Smith, "The History of the Tube Map," Londonist, April 2018, https://londonist.com/2016/05/the-history-of-the-tube-map.

92 from the Ancient Greek topos *("place, locality")*
"Topology," WordSense Online Dictionary, n.d., accessed 31 May 2023, https://www.wordsense.eu/topology/.

93 morphē *meaning "shape"*
"Homeomorphism (n.)," in Online Etymology Dictionary, n.d., accessed 31 May 2023, https://www.etymonline.com/word/homeomorphism.

93 its depiction on the Tube map
Matt Brown, "The Tube Map Is Full of Lies," Londonist, May 2017, https://londonist.com/london/transport/5-big-lies-on-the-tube-map.

94 after the 2014 Tube strike
Larcom, Rauch, and Willems, "Forced Experimentation."

95 it's a significantly slower route
Zhan Guo, "Mind the Map! The Impact of Transit Maps on Path Choice in Public Transit," *Transportation Research Part A: Policy and Practice* 45, no. 7 (2011): 625–639, https://doi.org/10.1016/j.tra.2011.04.001.

95 due to strike disruptions
Shaun Larcom, Ferdinand Rauch, and Tim Willems, "The Upside of London Tube Strikes," CentrePiece, Centre for Economic Performance, London School of Economics and Political Science, 6 November 2015, https://cep.lse.ac.uk/_new/publications/abstract.asp?index=4810.

95 preferring only the famous shape of Concorde
"Concorde Beats Tube Map by a Nose in UK Design Vote," Transport for London, 20 March 2006, https://tfl.gov.uk/info-for/media/press-releases/2006/march/concorde-beats-tube-map-by-a-nose-in-uk-design-vote.

96 gridded New York or San Francisco
Guo, "Mind the Map!"

96 Indians of the Catawba Nation
Tim St. Onge, "Celebrating Native American Cartography: The Catawba Deerskin Map," Library of Congress Blogs, 30 November 2016, https://blogs.loc.gov/maps/2016/11/celebrating-native-american-cartography-the-catawba-deerskin-map/.

97 the fourth- or fifth-century original
Oswald A.W. Dilke, "Itineraries and Geographical Maps in the Early and Late Roman Empires," in *The History of Cartography,* vol. 1, *Cartography in Prehistoric, Ancient, and Medieval Europe and the Mediterranean,* ed. John B. Harley and David Woodward (Chicago: University of Chicago Press, 1987), 234–257.

98 the island of Rhodes migrated next to Tel Aviv
Christos Nussli, "The Complete Tabula Peutingeriana: A Roman Road Map Compared with a Modern Map," Euratlas, September 2007, https://www.euratlas.net/cartogra/peutinger/index.html.

98 less than ideal for this purpose
"Legible London Yellow Book," Applied Information Group, 2007, https://content.tfl.gov.uk/ll-yellow-book.pdf.

99 carefully chosen positions within the city
"Legible London Yellow Book."

99 such a complicated layout
Derek Pollard, "Minutes, TfL and 'Legible London,'" Metric Views, 4 December 2018, https://metricviews.uk/2018/04/12/minutes-tfl-and-legible-london/.

99 long as half an hour to reach (see Figure 4.3)
Dan Saunders, "Crow Flies, Why Do People Still Use the Circle Based Approach?" Basemap blogs, 6 September 2017, https://web.archive.org/web/20171008040826/http://www.basemap.co.uk/crow-flies/.

99 farms, meadows, ponds, and marshes
Hilary Ballon, "Before the Grid," The Greatest Grid, Museum of the City of New York, 2015, https://thegreatestgrid.mcny.org/greatest-grid/before-the-grid.

100 forcing the state to offer them legal protection
Richard Kreitner, "First We Stake Manhattan," *Slate,* 5 April 2013, https://slate.com/culture/2013/04/marguerite-holloways-the-measure-of-manhattan-a-biography-of-john-randel-reviewed.html.

101 the inappropriate straight-line distance
Rizwan Shahid et al., "Comparison of Distance Measures in Spatial Analytical Modeling for Health Service Planning," *BMC Health Services Research* 9, no. 1 (2009), 200, https://doi.org/10.1186/1472-6963-9-200.

101 served students for about two thousand years
John Joseph O'Connor and Edmund Frederick Robertson, "Euclid of Alexandria," MacTutor (Maths History), School of Mathematics and Statistics, University of St Andrews, January 1999, https://mathshistory.st-andrews.ac.uk/Biographies/Euclid/.

105 gave them their current name
John Joseph O'Connor and Edmund Frederick Robertson, "René Maurice Fréchet," MacTutor (Maths History), School of Mathematics and Statistics, University of St Andrews, December 2005, https://mathshistory.st-andrews.ac.uk/Biographies/Frechet/.

105 headquarters in New Jersey
John Joseph O'Connor and Edmund Frederick Robertson, "Richard Wesley Hamming," MacTutor (Maths History), School of Mathematics and Statistics, University of St Andrews, January 2012, https://mathshistory.st-andrews.ac.uk/Biographies/Hamming/.

105 revolutionized information theory
Thomas M. Thompson, *From Error-Correcting Codes through Sphere Packings to Simple Groups* (Washington, DC: Mathematical Association of America, 1983), 17.

105 his method of error detection and error correction
Richard Wesley Hamming, "Error Detecting and Error Correcting Codes," *Bell System Technical Journal* 29, no. 2 (1950): 147–160, https://doi.org/10.1002/j.1538-7305.1950.tb00463.x.

108 They were right," Hamming recalled
Thompson, *Error-Correcting Codes,* 27.

108 receive satellite pictures from Mars
María Chara, "An Introduction to Error Correcting Codes," Mini Course 4, Mathematics sin Fronteras, Brown University, Lecture Notes 1, 12 May 2021, https://www.dam.brown.edu/MSF/Archives/Spring2021/notes/Course4Lecture1.pdf.

108 comparing DNA strings
Joseph Malkevitch, "Mathematics and the Genome: part 5. Near and Far (Strings)," Feature Column, American Mathematical Society, April 2002, https://www.ams.org/publicoutreach/feature-column/fcarc-genome5.

108 use QR codes
Wouter Boot, "QR Codes: What You Didn't Know," Acolad Group blog, 18 January 2022, https://blog.acolad.com/qr-codes-what-you-didnt-know#:~:text=Each%20QR%20code%20has%20one,%25%2C%20level%20H%2030%25.

108 constitute a positioning system in the brain
"The Nobel Prize in Physiology or Medicine 2014," The Nobel Prize, 28 December 2013, https://www.nobelprize.org/prizes/medicine/2014/summary/.

109 the rat moved to a different part of the room
May-Britt Moser and Noa Segev, "How Do We Find Our Way? Grid Cells in the Brain," Frontiers for Young Minds, 7 September 2021, https://kids.frontiersin.org/articles/10.3389/frym.2021.678725.

109 an accuracy of five centimeters
Moser and Segev, "How Do We Find Our Way?"

109 neighboring the hippocampus
Russell A. Epstein et al., "The Cognitive Map in Humans: Spatial Navigation and Beyond," *Nature Neuroscience* 20, no. 11 (2017): 1504–1513, https://doi.org/10.1038/nn.4656.

110 impact our estimation of the distance and direction
Michael Peer et al., "Structuring Knowledge with Cognitive Maps and Cognitive Graphs," *Trends in Cognitive Sciences* 25, no. 1 (2021): 37–54, https://doi.org/10.1016/j.tics.2020.10.004.

110 we're terrible at taking shortcuts
William H. Warren, "Non-Euclidean Navigation," *Journal of Experimental Biology* 222, suppl. 1 (2019), https://doi.org/10.1242/jeb.187971.

110 various measures as ninety degrees
Peer et al., "Structuring Knowledge."

111 topological map is blurry
Warren, "Non-Euclidean Navigation."

112 an average woman and an average man rather than individuals
Alexander P. Boone, Xinyi Gong, and Mary Hegarty, "Sex Differences in Navigation Strategy and Efficiency," *Memory & Cognition* 46, no. 6 (2018): 909–922, https://doi.org/10.3758/s13421-018-0811-y.

113 most likely cultural rather than biological
Hugo J. Spiers, Antoine Coutrot, and Michael Hornberger, "Explaining World-wide Variation in Navigation Ability from Millions of People: Citizen Science Project Sea Hero Quest," *Topics in Cognitive Science* 15, no. 1 (2021): 120–138, https://doi.org/10.1111/tops.12590.

5. Connected

114 about 30,000 pages in total

Javier Yanes, "Euler, the Beethoven of Mathematics," OpenMind, BBVA, 9 April 2018, https://www.bbvaopenmind.com/en/science/leading-figures/euler-the-beethoven-of-mathematics/.

114 Gdańsk in Poland

Roman Sznajder, "On Known and Less Known Relations of Leonhard Euler with Poland," *Studia Historiae Scientiarum* 15 (2016): 75–110, https://doi.org/10.4467/23921749shs.16.005.6148.

115 nobody had found such a path yet

Norman Biggs, Edward K. Lloyd, and Robin J. Wilson, *Graph Theory 1736–1936* (Oxford: Clarendon Press, 1986), 1–2.

115 does not depend on any mathematical principle

William Dunham, *The Genius of Euler: Reflections on His Life and Work* (Washington, DC: Mathematical Association of America, 2007), 266.

115 nor calculations made with them

Leonhard Euler, "Solutio Problematis Ad Geometriam Situs Pertinentis," *Commentarii Academiae Scientiarum Imperialis Petropolitanae* 8 (1741): 128–140.

116 Euler stripped Königsberg's map of its distracting details

Biggs, Lloyd, and Wilson, *Graph Theory*, 8–9.

119 French mathematician Pierre-Henry Fleury

Pierre-Henry Fleury, "Deux problèmes de Géométrie de situation," *Journal de mathématiques élémentaires* 2, no. 2 (1883): 257–261.

121 study abstract mathematical structures

Biggs, Lloyd, and Wilson, *Graph Theory*, 9.

121 25.2 million packages

UPS, "Notice of 2022 Annual Meeting of Shareowners and Proxy Statement & 2021 Annual Report on Form 10-K," 19, 21 March 2019, https://investors.ups.com/_assets/_132ec38659b7c1b3c59113edc15ef905/ups/db/1175/10620/annual_report/UPS_2022_Proxy_Statement_and_2021_Annual_Report%3B_Form_10-K.pdf.

121 135 times a day

"UPS to Enhance ORION with Continuous Delivery Route Optimization," press release, UPS Newsroom, 29 January 2020, https://about.ups.com/ae/en/newsroom/press-releases/innovation-driven/ups-to-enhance-orion-with-continuous-delivery-route-optimization.html#:~:text=ORION%20is%20a%20proprietary%20technology,About%20UPS.

122 an important profession

Jakob Tanner, "Society & Stigma: Brief History of Salesmen," LONDNR, 4 December 2016, https://web.archive.org/web/20210624234942/https://www.londnr.com/brief-history-of-salesmen/.

122 Old Traveling Salesman
Alexander Schrijver, "On the History of Combinatorial Optimization (Till 1960)," in *Discrete Optimization,* ed. Karen Aardal, George L. Nemhauser, and Robert Weismantel (Amsterdam: Elsevier, 2005), 1–68.

122 first recorded mention of the route planning problem
William J. Cook, *In Pursuit of the Traveling Salesman: Mathematics at the Limits of Computation* (Princeton, NJ: Princeton University Press, 2011), 19–24.

122 without having to touch the same place twice
Schrijver, "On the History of Combinatorial Optimization," 49.

123 find the shortest path connecting the points
Cook, *In Pursuit,* 36.

123 support the US Armed Forces
"A Brief History of RAND," RAND Corporation, Santa Monica, CA, n.d., accessed 30 May 2023, https://www.rand.org/about/history.html.

125 4.5 times the shortest one
Cook, *In Pursuit,* 65–68.

126 researchers developed another clever heuristic
John J. Bartholdi III et al., "A Minimal Technology Routing System for Meals on Wheels," *Interfaces* 13, no. 3 (1983): 1–8, https://doi.org/10.1287/inte.13.3.1.

128 it picks a direction at random
Marco Dorigo and Luca Maria Gambardella, "Ant Colonies for the Travelling Salesman Problem," *Biosystems* 43, no. 2 (1997): 73–81, https://doi.org/10.1016/s0303-2647(97)01708-5.

128 bios *("life") and* mīmēsis *("imitation")*
Julian F. V. Vincent et al., "Biomimetics: Its Practice and Theory," *Journal of the Royal Society: Interface* 3, no. 9 (2006): 471–482, https://doi.org/10.1098/rsif.2006.0127.

129 bankruptcy prediction
Yudong Zhang, Shuihua Wang, and Genlin Ji, "A Rule-Based Model for Bankruptcy Prediction Based on an Improved Genetic Ant Colony Algorithm," *Mathematical Problems in Engineering* 2013 (2013): 753251, https://doi.org/10.1155/2013/753251.

129 power electronic circuit design
Jun Zhang et al., "Extended Ant Colony Optimization Algorithm for Power Electronic Circuit Design," *IEEE Transactions on Power Electronics* 24, no. 1 (2009): 147–162, https://doi.org/10.1109/tpel.2008.2006175.

129 responsibility of one driver
Chuck Holland et al., "UPS Optimizes Delivery Routes," *Interfaces* 47, no. 1 (2017): 8–23, https://doi.org/10.1287/inte.2016.0875.

133 no two regions sharing a boundary have the same color
Robin Wilson, *Four Colors Suffice,* rev. ed. (Princeton, NJ: Princeton University Press, 2013), 16–21.

134 like the five-country "map" in Figure 5.11
Heinrich Tietze, *Famous Problems of Mathematics* (New York: Graylock Press, 1966), 77–78.

135 I think I must do as the Sphynx did . . .
Wilson, *Four Colors,* 18.

138 it was a very good incorrect proof
Wilson, *Four Colors,* 116.

139 vertices with more than four neighbors
kabenyuk, post to "How to construct a planar graph (or a class of planar graphs) with minimum degree 5 of diameter 2?" Mathematics Stack Exchange, 9 November 2022, 12:02, https://math.stackexchange.com/questions/4572624/how-to-construct-a-planar-graph-or-a-class-of-planar-graphs-with-minimum-degre.

141 "We now know that this policy was essential to our success."
Kenneth Appel and Wolfgang Haken, "The Solution of the Four-Color-Map Problem," *Scientific American* 237, no. 4 (1977): 108–121, https://doi.org/10.1038/scientificamerican1077-108.

142 It is also conceivable that no such proof is possible
Appel and Haken, "The Solution."

143 started teaching future mathematicians to use them
Emily Riehl, course website for Math 301: Introduction to Proofs, Johns Hopkins University, Spring 2019, https://math.jhu.edu/~eriehl/301-s19/.

6. Divided

144 in Asheville, North Carolina, in November 2017
Laurel Wamsley, "Not-So-Fun Run: Joggers in 'Gerrymander 5K' Must Run Oddly Shaped Route," NPR, 30 October 2017, https://www.npr.org/sections/thetwo-way/2017/10/30/560909678/not-so-fun-run-gerrymander-5k-joggers-trace-asheville-s-electoral-district.

144 "Brooklyn of North Carolina"
Kristin Hunt and Lee Breslouer, "The Brooklyn of Every State," Thrillist, 25 July 2015, https://www.thrillist.com/entertainment/nation/hipster-neighborhoods-the-brooklyn-of-every-state.

145 the League of Women Voters
"About Us," League of Women Voters, n.d., accessed 1 June 2023, https://www.lwv.org/about-us.

145 the curious 5K boundary
"North Carolina Residents Protest Skewed Electoral Boundaries with 'Gerrymander 5K' Race," CBC Radio, 1 November 2017, https://www.cbc.ca/radio/asithappens/as-it-happens-tuesday-edition-1.4373278/north-carolina-residents-protest-skewed-electoral-boundaries-with-gerrymander-5k-race-1.4380391.

147 leaves the remaining seats for the incumbent, like in Figure 6.2d
Olga Pierce, Jeff Larson, and Lois Beckett, "Redistricting, a Devil's Dictionary," ProPublica, 2 November 2011, https://www.propublica.org/article/redistricting-a-devils-dictionary.

147 Asheville and its conservative suburbs
"North Carolina Congressional Redistricting after the 2010 Census," Carolina Demography, UNC Population Center, University of North Carolina at Chapel Hill, n.d., accessed 1 June 2023, https://www.arcgis.com/apps/StorytellingSwipe/index.html?appid=a15c27c984ed404782da753dd840e99a#.

147 in either District 10 or District 11
Jeremy Markovich, "I Ran the Worst 5K of My Life So I Could Explain Gerrymandering to You," Politico, 15 November 2017, https://www.politico.com/magazine/story/2017/11/15/gerrymandering-5k-asheville-north-carolina-215829/.

147 he came away none the wiser
Markovich, "Worst 5K."

147 some votes would matter more than others
Wesberry v. Sanders, 376 US1 (1964) Oyez, n.d., accessed 1 June 2023, https://www.oyez.org/cases/1963/22.

148 plus or minus one voter
NC 2022 Congressional, map, Dave's Redistricting, accessed 1 June 2023, https://davesredistricting.org/maps#viewmap::6e8268a4-3b9b-4140-8f99-e3544a2f0816.

148 as long as the numbers added up
Markovich, "Worst 5K."

148 kept by the Library of Congress
"Original Woodblocks for Printing 'Gerrymander' Political Cartoon Map That Was Issued in Boston Gazette of March 26, 1812," Library of Congress, LCCN 2003620165, https://lccn.loc.gov/2003620165.

148 over eighty newspapers all over the United States
Nicholas R. Seabrook, *One Person, One Vote: A Surprising History of Gerrymandering in America* (New York: Pantheon Books, 2022).

148 reaching 100,000 in the early nineteenth century
Emma Griffin, "Manchester in the 19th Century," British Library, 15 May 2015, https://www.bl.uk/romantics-and-victorians/articles/manchester-in-the-19th-century.

149 two representatives in Parliament
Seabrook, *One Person, One Vote.*

149 changed the electoral system of England and Wales
Seabrook, *One Person, One Vote.*

149 (Madison won anyway)
Erik J. Engstrom, *Partisan Gerrymandering and the Construction of American Democracy* (Ann Arbor: University of Michigan Press, 2013).

149 to disadvantage an incumbent party's opponents
Bernard Grofman and German Feierherd, "The U.S. Could Be Free of Gerrymandering. Here's How Other Countries Do Redistricting," *Washington Post,* 7 August 2017.

149 the small populations of these islands
Chloe Smith, "Update: Strengthening Democracy," UK Parliament Statement, 24 March 2020, https://questions-statements.parliament.uk/written-statements/detail/2020-03-24/HCWS183.

149 the added communities would get thirty-two
"Polish Opposition Accuses Ruling Party of Power Grab in Warsaw," Reuters, 2 February 2017, https://www.reuters.com/article/uk-poland-politics-warsaw-idUKKBN15H20P.

150 packed the opposition voters into just a few large districts
Kim Lane Scheppele, "How Viktor Orbán Wins," *Journal of Democracy* 33, no. 3 (July 2022): 45–61, https://doi.org/10.1353/jod.2022.0039.

151 results rather than district shapes to detect gerrymandering
Nicholas O. Stephanopoulos and Eric M. McGhee, "Partisan Gerrymandering and the Efficiency Gap," *University of Chicago Law Review* 82, no. 2 (2015): 831–900.

151 gerrymandering cases: the efficiency gap
Michael Wines, "Judges Find Wisconsin Redistricting Unfairly Favored Republicans," *New York Times,* 21 November 2016.

152 wasted the same number of votes
Stephanopoulos and McGhee, "Partisan Gerrymandering," 834.

153 the efficiency gap penalizes proportionality
Mira Bernstein and Moon Duchin, "A Formula Goes to Court: Partisan Gerrymandering and the Efficiency Gap," *Notices of the American Mathematical Society* 64, no. 9 (2017): 1020–1024, https://doi.org/10.1090/noti1573.

155 town or county lines, whenever possible
"Where Are the Lines Drawn?" All about Redistricting, Loyola Law School, 2 April 2021, https://redistricting.lls.edu/redistricting-101/where-are-the-lines-drawn/.

155 creating urban–rural political divides
Rahsaan Maxwell, "Why Are Urban and Rural Areas So Politically Divided?" *Washington Post,* 5 March 2019.

156 the world's fastest supercomputers
Yan Y. Liu, Wendy K. Tam Cho, and Shaowen Wang, "PEAR: A Massively Parallel Evolutionary Computation Approach for Political Redistricting Optimization and Analysis," *Swarm and Evolutionary Computation* 30 (2016): 78–92, https://doi.org/10.1016/j.swevo.2016.04.004.

156 the number of cells in a human body
Moon Duchin, "Graphs, Geometry and Gerrymandering," Gathering for Gardner (G4G) Celebration of Mind, 23 October 2021, YouTube video, https://www.youtube.com/watch?v=VU8CtVmiP3w.

156 way beyond our reach
U.S. Election Assistance Commission to the 117th Congress, *Election Administration and Voting Survey 2020 Comprehensive Report,* August 2021, 43, https://www.eac.gov/sites/default/files/document_library/files/2020_EAVS_Report_Final_508c.pdf.

157 is called "random seed and grow"
Mike Orcutt, "How Math Has Changed the Shape of Gerrymandering," Quanta Magazine, 1 June 2023, https://www.quantamagazine.org/how-math-has-changed-the-shape-of-gerrymandering-20230601/.

157 deeming any comparisons with the real map misleading
Wendy K. Tam Cho and Yan Y. Liu, "Sampling from Complicated and Unknown Distributions," *Physica A: Statistical Mechanics and Its Applications* 506 (15 September 2018): 170–178, https://doi.org/10.1016/j.physa.2018.03.096

157 animals or plants over time
Liu, Cho, and Wang, "PEAR."

157 called Markov chain Monte Carlo (MCMC)
Benjamin Fifield et al., "Automated Redistricting Simulation Using Markov Chain Monte Carlo," *Journal of Computational and Graphical Statistics* 29, no. 4 (2020): 715–728, https://doi.org/10.1080/10618600.2020.1739532.

158 led by mathematician Moon Duchin
Daryl DeFord, Moon Duchin, and Justin Solomon, "Recombination: A Family of Markov Chains for Redistricting," *Harvard Data Science Review* 3, no. 1 (March 31, 2021), https://doi.org/10.1162/99608f92.eb30390f.

158 a bipartisan group of voters
Wendy K. Tam Cho and Yan Y. Liu, "Toward a Talismanic Redistricting Tool: A Computational Method for Identifying Extreme Redistricting Plans," *Election Law Journal: Rules, Politics, and Policy* 15, no. 4 (2016): 351–366, https://doi.org/10.1089/elj.2016.0384.

160 three factors that must be satisfied
"Majority-Minority Districts," Ballotpedia, n.d., accessed 1 June 2023, https://ballotpedia.org/Majority-minority_districts.

161 printed on his and his wife's wedding cake
Robert Gebelhoff, "Think You Hate Gerrymandering? Think Again," *Washington Post,* 11 May 2018, opinion video, https://www.youtube.com/watch?v=eomnA9zZT94.

161 took the ambulance over an hour to arrive
"Utah's Navajo Residents Hope Redistricting Brings Needed Resources," PBS News Hour, 21 June 2018, https://www.pbs.org/newshour/show/utahs-navajo-residents-hope-redistricting-brings-needed-resources.

161 closer to the reservation
Krista Langlois, "How a Utah County Silenced Native American Voters—and How Navajos Are Fighting Back," *High Country News,* 13 June 2016, https://www.hcn.org/issues/48.10/how-a-utah-county-silenced-native-american-voters-and-how-navajos-are-fighting-back.

162 with two-thirds of the population being Native Americans
Matthew Isbell, "San Juan County, Utah Maintains Fair Commission Maps for the Navajo," MCI Maps, 7 February 2022, https://mcimaps.com/san-juan-county-utah-maintains-fair-commission-maps-for-the-navajo/.

162 the majority vote in the county commission
Kate Groetzinger, "In Utah, Change Is Slow Following Historic Election of Native Americans," The GroundTruth Project, 20 February 2020, https://thegroundtruthproject.org/in-utah-change-is-slow-following-historic-election-of-native-americans/.

163 this oil-rich region
Henry Grabar, "The Battle for San Juan County, Utah," *Slate,* 25 August 2020, https://slate.com/news-and-politics/2020/08/san-juan-county-utah-native-americans-republicans-bears-ears.html.

163 don't have access to clean running water
Kate Groetzinger, "In Utah."

163 have already done this
"Redistricting Commissions," Ballotpedia, n.d., accessed 1 June 2023, https://ballotpedia.org/Redistricting_commissions.

163 who tend to live together by choice
Christine Laskowski and Galen Druke, "Is Gerrymandering My Fault?" Gerrymandering Project, FiveThirtyEight, 23 January 2018, video, https://fivethirtyeight.com/videos/is-gerrymandering-my-fault/.

163 proportional share of seats
Christopher S. Fowler and Linda L. Fowler, "Here's a Different Way to Fix Gerrymandering," *Washington Post,* 6 July 2021.

164 seats would become more balanced
Jonathan Rodden, interview by Wioletta Dziuda, Anthony Fowler, and William Howell, "Why Democrats Should Move to the Suburbs If They Want to Win More Legislative Seats," Not Another Politics Podcast, episode 28, Harris School of Public Policy, University of Chicago, 7 April 2021, https://harris.uchicago.edu/news-events/news/why-democrats-should-move-suburbs-if-they-want-win-more-legislative-seats.

164 drawing large districts to dilute Black votes
Strategy 1, Recommendation 3, "Amend or repeal and replace the 1967 law . . . ," in "Our Common Purpose: Reinventing American Democracy for the 21st Century," report of Commission on the Practice of Democratic Citizenship, American Academy of Arts and Sciences, June 2020, https://www.amacad.org/ourcommonpurpose/report.

164 which eliminates the least-preferred candidates
"Ranked-Choice Voting (RCV)," Ballotpedia, n.d., accessed 1 June 2023, https://ballotpedia.org/Ranked-choice_voting_(RCV).

164 proportional representation of minority groups
Gerdus Benade et al., "Ranked Choice Voting and Minority Representation," unpublished manuscript, 18 February 2021, last revised 14 July 2022, https://doi.org/10.2139/ssrn.3778021.

165 their children were currently assigned
Alvin Chang, "School Segregation Didn't Go Away. It Just Evolved," Vox, 27 July 2017, https://www.vox.com/policy-and-politics/2017/7/27/16004084/school-segregation-evolution.

165 close to a renowned school
Tommy Unger, "Affording a House in a Highly Ranked School Zone? It's Elementary," Redfin Real Estate News, 25 September 2013, updated 6 October 2020, https://www.redfin.com/news/paying-more-for-a-house-with-a-top-public-school-its-elementary/.

165 he filed a lawsuit, which led to the historical success
Brian Duignan, "Brown v. Board of Education," *Encyclopaedia Britannica*, last updated 10 May 2023, https://www.britannica.com/event/Brown-v-Board-of-Education-of-Topeka.

165 three-quarters of students are either white or nonwhite
"Nonwhite School Districts Get $23 Billion Less Than White Districts," EdBuild, February 2019, https://edbuild.org/content/23-billion.

165 which adds up to a $23 billion gap
"Nonwhite School Districts."

165 then go home
Emma Brown, "Judge: Mostly White Southern City May Secede from School District despite Racial Motive," *Washington Post,* 27 April 2017.

166 less privileged students left behind
"Fractured: The Accelerating Breakdown of America's School Districts, Case Study: Jefferson County, AL," EdBuild, 2019, https://edbuild.org/content/fractured#jeffco.

166 five thousand residents can create a separate school district
"Fractured."

166 thus forbidding the secession
"Fractured."

167 which created de facto ghettoes within cities
Chang, "School Segregation Didn't Go Away."

167 most students were native Danish speakers
Anna Piil Damm et al., "Effects of Busing on Test Scores and the Wellbeing of Bilingual Pupils: Resources Matter," Economics Working Papers 2020–03, Department of Economics and Business Economics, Aarhus University, May 2020, https://ideas.repec.org/p/aah/aarhec/2020-03.html.

167 countries as diverse as Finland, Chile, and Australia
Venla Bernelius and Katja Vilkama, "Pupils on the Move: School Catchment Area Segregation and Residential Mobility of Urban Families," *Urban Studies* 56, no. 15 (2019): 3095–3116, https://doi.org/10.1177/0042098019848999; Claudio Allende González, Rocío Díaz, and Juan Pablo Valenzuela, "School Segregation in Chile," in *Global Encyclopedia of Public Administration, Public Policy, and Governance,* ed. Ali Farazmand (Cham, Switzerland: Springer, 2018), 5544–5554; Emma E. Rowe and Christopher Lubienski, "Shopping for Schools or Shopping for Peers: Public Schools and Catchment Area Segregation," *Journal of Education Policy* 32, no. 3 (2016): 340–356, https://doi.org/10.1080/02680939.2016.1263363.

167 priority in the eight most local schools
"Primary School" City of Amsterdam, accessed 4 February 2024, https://www.amsterdam.nl/en/education/primary-school/#:~:text=Your%20child%20will%20have%20priority,then%20your%20third%20choice%2C%20etc.

167 they tend to select the closest institution
Tracy Brown Hamilton, "Solving the Problem of Amsterdam's 'Black' and 'White' Schools," *The Atlantic,* 18 June 2015.

168 this school is disproportionately white
Tomás Monarrez et al., "How Much Does Your School Contribute to Segregation?" Urban Institute, 8 July 2020, https://apps.urban.org/features/school-segregation-index/.

168 developed the Segregation Contribution Index (SCI)
Monarrez et al., "How Much Does Your School Contribute?"

168 exactly reflected the district's composition
Tomás Monarrez, Brian Kisida, and Matthew Chingos, "When Is a School Segregated? Making Sense of Segregation 65 Years after Brown v. Board of Education," report, Urban Institute, 27 September 2019, https://www.urban.org/research/publication/when-school-segregated-making-sense-segregation-65-years-after-brown-v-board-education.

169 over 100,000 in all
"Index of School Contribution to the Racial Segregation of US School Districts," Data Catalog, Urban Institute, release date 25 June 2020, accessed 8 July 2020, https://datacatalog.urban.org/dataset/index-school-contribution-racial-segregation-us-school-districts.

169 school segregation within the district
Monarrez, Kisida, and Chingos, "When Is a School Segregated?"

169 segregation levels seem to have worsened
Deborah Wilson and Gary Bridge, "School Choice and the City: Geographies of Allocation and Segregation," *Urban Studies* 56, no. 15 (2019): 3198–3215, https://doi.org/10.1177/0042098019843481.

170 south of the original boundary
Chang, "School Segregation Didn't Go Away."

170 to find the most segregating ones

Tomás Monarrez and Carina Chien, "Dividing Lines: Racially Unequal School Boundaries in US Public School Systems," report, Urban Institute, 1 September 2021, https://www.urban.org/research/publication/dividing-lines-racially-unequal-school-boundaries-us-public-school-systems.

171 optimize catchment areas in Swiss cities

Oliver Dlabac, Adina Amrhein, and Fabienne Hug, "School Mixing: More Equity through Intelligent School Zoning," Study Report No. 17, Center for Democracy Studies Aarau, Switzerland, April 2022, https://villejuste.com/wp-content/uploads/2022/04/School_mixing_intelligent_school_zoning_report.pdf.

171 the way states fund public schools

EdBuild homepage, accessed 1 June 2023, https://edbuild.org/.

171 without any raises in taxes or debt

"Clean Slate," EdBuild, May 2020, https://edbuild.org/content/clean-slate.

172 advice on fighting school segregation

"European Guidelines," European Cities against School Segregation (ECASS), n.d., accessed 1 June 2023, https://www.ecass.eu/european-guidelines/.

7. Found

173 In 1831, cholera arrived in England

Max Roser et al., "Eradication of Diseases," Our World in Data, June 2014, rev. October 2018, https://ourworldindata.org/eradication-of-diseases.

173 it leads to death, often within hours

Miriam Reid, "John Snow Hunts the Blue Death," *Distillations,* Science History Institute, Philadelphia, 8 March 2022, https://www.sciencehistory.org/distillations/john-snow-hunts-the-blue-death.

173 a mining village close to Newcastle

Reid, "John Snow Hunts."

173 instead of ingesting poison

John Snow, *On the Mode of Communication of Cholera* (London: John Churchill, 1849), 6–7.

173 London's third cholera outbreak in 1854

E. Ashworth Underwood, "The History of Cholera in Great Britain," *Proceedings of the Royal Society of Medicine* 41, no. 3 (March 1948): 165–173, https://doi.org/10.1177/003591574804100309.

174 modes of transmission of cholera

Kenneth Field, "Something in the Water: The Mythology of Snow's Map of Cholera," ArcGIS blog, Environmental Systems Research Institute (ESRI), Redlands, CA, 3 December 2020, https://www.esri.com/arcgis-blog/products/arcgis-pro/mapping/something-in-the-water-the-mythology-of-snows-map-of-cholera/.

174 the disease had taken dozens of victims
Snow, *On the Mode of Communication of Cholera* (1849), 12–23.

174 Snow set out to find a pattern
"England: The Broad Street Pump—Epidemiology Begins!" part 2, Extra History YouTube channel, 21 November 2015, https://www.youtube.com/watch?v=1jlsyucUwpo&list=PLhyKYa0YJ_5Aq7g4bil7bnGi0A8gTsawu&index=52.

174 only 37 by the latter?
John Snow, *On the Mode of Communication of Cholera,* 2nd ed., much enlarged (London: John Churchill, 1855), 86.

174 which makes it safe for the environment
"Sewage Treatment," River Thame Conservation Trust, 2018, https://web.archive.org/web/20230428104606/https://riverthame.org/river-thame/pressures/water-quality/sewage-treatment/.

174 "in a most impure condition"
Snow, *On the Mode of Communication of Cholera* (1855), 64.

175 possibly the first ever natural experiment
Peter Craig et al., "Natural Experiments: An Overview of Methods, Approaches, and Contributions to Public Health Intervention Research," *Annual Review of Public Health* 38, no. 1 (2017): 39–56, https://doi.org/10.1146/annurev-publhealth-031816-044327.

176 the water of the different Companies
Snow, *On the Mode of Communication of Cholera* (1855), 75.

176 in most cases, without their knowledge
Snow, *On the Mode of Communication of Cholera* (1855), 75.

177 Golden Square in central London's Soho
Steven Johnson, *The Ghost Map: The Story of London's Most Terrifying Epidemic—and How It Changed Science, Cities, and the Modern World* (New York: Riverhead Books, 2006), 109.

177 a map of the Broad Street area
Field, "Something in the Water."

179 preventing forest fires
Lifeng Liu et al., "The Application of Voronoi Algorithm in the Planning of Forest-Fire," *IOP Conference Series: Materials Science and Engineering* 490, no. 4 (2019): 042008, https://doi.org/10.1088/1757-899x/490/4/042008.

179 understanding patterns on animal coats
Marcelo Walter, Alain Fournier, and Daniel Menevaux, "Integrating Shape and Pattern in Mammalian Models," *Proceedings of the 28th Annual Conference on Computer Graphics and Interactive Techniques* (New York: Association for Computing Machinery, 2001), 317–326, https://doi.org/10.1145/383259.383294.

179 the Broad Street well water
Johnson, *Ghost Map,* 160.

179 Soon, Soho was cholera-free
Underwood, "History of Cholera."

179 trying to understand what had happened
Johnson, *Ghost Map,* 163.

179 where the contaminant of the Broad Street pump came from
Johnson, *Ghost Map,* 172–181.

179 gotten the symptoms around August 28
Johnson, *Ghost Map,* 177.

180 in the original location
"Welcome to the John Snow Society," The John Snow Society, London, 5 December 2022, https://johnsnowsociety.org/#The-Society.

180 among his contemporaries and even today
Tom Koch and Kenneth Denike, "Crediting His Critics' Concerns: Remaking John Snow's Map of Broad Street Cholera, 1854," *Social Science and Medicine* 69, no. 8 (October 2009): 1246–1251, https://doi.org/10.1016/j.socscimed.2009.07.046.

180 future improvements in sanitation
"Cholera in Victorian London," Science Museum, London, 30 July 2019, https://www.sciencemuseum.org.uk/objects-and-stories/medicine/cholera-victorian-london.

180 based on his studies of typhoid fever
Robert Moorhead, "William Budd—a Less Well Known Human Ecologist?" commentary, *International Journal of Epidemiology* 42, no. 6 (December 2013): 1578–1579, https://doi.org/10.1093/ije/dyt222.

181 the yellow fever outbreak in New York City
Tom Koch, "Mapping the Miasma: Air, Health, and Place in Early Medical Mapping," *Cartographic Perspectives,* no. 52 (September 1, 2005): 4–27, https://doi.org/10.14714/cp52.376.

181 a recent influenza epidemic
Sarah Hepworth, "Book of the Month": Robert Perry, "Facts and Observations on the Sanitary State of Glasgow," (1844), Glasgow University Library Special Collections Department, February 2006, https://www.gla.ac.uk/myglasgow/library/files/special/exhibns/month/feb2006.html.

181 mapping spatial data had already gone mainstream
Koch and Denike, "Crediting His Critics' Concerns."

181 areas close to other pumps
Koch and Denike, "Crediting His Critics' Concerns."

182 became popular in the early twentieth century
Fred Brauer, "Compartmental Models in Epidemiology," *Mathematical Epidemiology* 1945 (2008): 19–79, https://doi.org/10.1007/978-3-540-78911-6_2.

182 removed (sometimes called recovered, R)
Ana Pastore y Piontti et al., *Charting the Next Pandemic: Modeling Infectious Disease Spreading in the Data Science Age* (Cham, Switzerland: Springer, 2019), 35–37.

183 real-world population and mobility data
Pastore y Piontti et al., *Charting the Next Pandemic,* 29–34.

185 enroll in one in the future
Allan Casey, "Rossmo's Formula," College of Arts and Science, University of Saskatchewan, 8 May 2018, https://artsandscience.usask.ca/magazine/Spring_2018/rossmos-formula.php.

185 kept his job there for almost twenty-three years
Casey, "Rossmo's Formula."

186 called a Doctor of Criminology
Casey, "Rossmo's Formula."

186 socioeconomic characteristics of the criminal
D. Kim Rossmo, "Recent Developments in Geographic Profiling," *Policing: A Journal of Policy and Practice* 6, no. 2 (9 June 2012): 144–150, https://doi.org/10.1093/police/par055.

186 the same place at the same time
João Medeiros, "How Geographic Profiling Helps Find Serial Criminals," *Wired,* 18 November 2014, https://www.wired.co.uk/article/mapping-murder.

186 introduce other preventative measures
Patricia Brantingham and Paul Brantingham, "Criminality of Place: Crime Generators and Crime Attractors," *European Journal on Criminal Policy and Research* 3, no. 3 (1995): 1–26, https://doi.org/10.1007/bf02242925.

187 we can figure out the sprinkler's position
Medeiros, "How Geographic Profiling Helps."

187 returns there afterward
Environmental Criminology Research Inc., "Introduction to Geographic Profiling for Crime Analysis," PowerPoint presentation, Vancouver, BC, 2012.

189 became infamous as the "chair burglar"
D. Kim Rossmo and Lorie Velarde, "Geographic Profiling Analysis: Principles, Methods and Applications," in *Crime Mapping Case Studies: Practice and Research,* ed. Spencer Chainey and Lisa Tompson (Chichester: John Wiley and Sons, 2008), 35–43.

189 Lorie Velarde gets straight to work
"Lorie Velarde, Geographic Profiling Analyst," Analyst Talk with Jason Elder, 20 June 2022, Law Enforcement Analysis Podcasts, no. 00112, https://www.leapodcasts.com/e/atwje-lorie-velarde-the-geographic-profiler/.

190 a property loss of over $2.5 million
Rossmo and Velarde, "Geographic Profiling Analysis,"41–42.

190 but also to counter terrorism

Henry Kucera, "Hunting Insurgents: Geographic Profiling Adds a New Weapon," *GeoWorld* 18, no. 10 (2005): 30–32.

190 pirate ship bases in the Caribbean

Michał Górski, "Geographic Profiling and Event Prediction for Seventeenth-and Eighteenth-Century Pirates," *Professional Geographer* 74, no. 4 (2022): 781–791, https://doi.org/10.1080/00330124.2022.2075405.

190 determine earthquake epicenters from digital reports

Aurélien Dupont et al., "Determination of the Earthquake Epicenter from the Geographic Profiling of the Digital Footprints Left by Eyewitnesses," poster presented at the 36th General Assembly of the European Seismological Commission, Valetta, Malta, 2–7 September 2018, https://www.emsc.eu/Files/docs/publications/Poster_ESC2018-S17-205.pdf.

190 hunting patterns of white sharks

R. A. Martin, D. K. Rossmo, and N. Hammerschlag, "Hunting Patterns and Geographic Profiling of White Shark Predation," *Journal of Zoology* 279, no. 2 (2009): 111–118, https://doi.org/10.1111/j.1469-7998.2009.00586.x.

193 clinging to their initial idea

Colleen Keller Sterling, "Bayesian Search for Missing Aircraft," 2021 MORS talks, episode 47, 7 May 2021, Military Operations Research Society, Arlington, VA, YouTube video, https://www.youtube.com/watch?v=6yghqIUcUcM.

195 multiple countries and organizations

Kirk Semple, "Search for Malaysian Jet to Be Costliest in History," *New York Times,* 9 April 2014, sec. A.

196 receiving the information about the crash

Lawrence D. Stone, "In Search of Air France Flight 447," *OR/MS Today* 38, no. 4 (2011), Institute for Operations Research and Management Science, https://www.informs.org/ORMS-Today/Public-Articles/August-Volume-38-Number-4/In-Search-of-Air-France-Flight-447.

196 which describes motion in water

Stone, "In Search of Air France Flight 447."

197 determining the causes of the crash

Lawrence D. Stone et al., "Search for the Wreckage of Air France Flight AF 447," *Statistical Science* 29, no. 1 (February 2014): 69–80, https://doi.org/10.1214/13-sts420.

198 plunge into the water at a breathtaking speed

William Langewiesche, "The Human Factor," *Vanity Fair,* 17 September 2014.

198 controling the plane by hand

Angelique Chrisafis, "Final Minutes of Air France Flight AF447 to Be Examined as Trial Opens," *The Guardian,* 10 October 2022.

8. Deep

200 the surface of Mars
Daniel O'Donohue, "Mapping the Ocean Floor," Mapscaping podcasts, 21 January 2021, https://mapscaping.com/podcast/mapping-the-ocean-floor/.

200 reflected from the Earth's surface
"Altimetric Bathymetry," NOAA / NESDIS / STAR, Laboratory for Satellite Altimetry, n.d., accessed 2 June 2023, https://www.star.nesdis.noaa.gov/socd/lsa/AltBathy/#:~:text=Satellite.

201 crossing the oceans every day
Chris Baraniuk, "What It's Like to Sail a Giant Ship on Earth's Busiest Seas," BBC, 28 November 2016, https://www.bbc.com/future/article/20161128-what-its-like-to-sail-colossal-ships-on-earths-busiest-sea#:~:text=On.

201 know what lies on the bottom
Jess I. T. Hillman, "Mapping the Oceans," Frontiers for Young Minds, 19 February 2019, https://kids.frontiersin.org/articles/10.3389/frym.2019.00025.

201 mapping the entire ocean floor by 2030
"Our Mission," The Nippon Foundation-GEBCO Seabed 2030, accessed 4 February 2024, https://seabed2030.org/our-mission/.

201 the oceans, seas, and marine resources
"Goal 14: Conserve and Sustainably Use the Oceans, Seas and Marine Resources," 17 Sustainable Development Goals, United Nations, adopted September 2015, https://www.un.org/sustainabledevelopment/oceans/.

202 a weight attached to one end
Anne-Cathrin Wölfl et al., "Seafloor Mapping—the Challenge of a Truly Global Ocean Bathymetry," *Frontiers in Marine Science* 6 (2019), art. 283, 2, https://doi.org/10.3389/fmars.2019.00283.

202 detecting underwater objects
Wölfl et al., "Seafloor Mapping," 2.

202 receives back a delayed signal
Mark Monmonier, "Hydrographic Techniques," in *The History of Cartography*, vol. 6, *Cartography in the Twentieth Century*, ed. Mark Monmonier (Chicago: University of Chicago Press, 2015).

203 level of detail included in the map
"How Multibeam Sonar Works," NOAA Ocean Explorer, 2009, https://oceanexplorer.noaa.gov/explorations/09bermuda/background/multibeam/multibeam.html.

203 10 meters apart along a width of about 9,000 meters
"Acoustic Systems," Science & Tech, Nautilus Live, n.d., accessed 1 June 2023, https://nautiluslive.org/tech/acoustic-systems.

204 to increase the resolution
"How Multibeam Sonar Works."

204 reflect more sound than a bumpier seafloor
"How Does Backscatter Help Us Understand the Sea Floor?" NOAA National Ocean Service, 2 April 2019, last updated 4 October 2023, https://oceanservice.noaa.gov/facts/backscatter.html.

204 a specialized ship, Nautilus
Nautilus Live, accessed 1 June 2023, https://nautiluslive.org/.

205 Papahānaumokuākea Marine National Monument
"Ala 'Aumoana Kai Uli in Papahānaumokuākea Marine National Monument," NA 154 Expedition, September 2023, Nautilus Live, n.d., accessed 1 June 2023, https://nautiluslive.org/cruise/na154.

205 student at the University of Ohio
Marie Tharp, "Connect the Dots: Mapping the Seafloor and Discovering the Mid-Ocean Ridge," in *Lamont-Doherty Earth Observatory: Twelve Perspectives on the First Fifty Years, 1949–1999,* ed. Laurence Lippsett (Palisades, NY: Lamont-Doherty Earth Observatory of Columbia University, 1999).

206 the guys were off fighting
Tharp, "Connect the Dots."

206 from the University of Tulsa
Marie Tharp and Henry Frankel, "Mappers of the Deep," *Natural History* 95, no. 10 (October 1986).

206 which they left to their female colleagues
Naomi Oreskes, "Laissez-tomber: Military Patronage and Women's Work in Mid-20th-Century Oceanography," *Historical Studies in the Physical and Biological Sciences* 30, no. 2 (January 2000): 373–392, https://doi.org/10.2307/27757836.

206 putting the data into a geographical context
Betsy Mason, "Marie Tharp's Groundbreaking Maps Brought the Seafloor to the World," Science News, 13 January 2021, https://www.sciencenews.org/article/marie-tharp-maps-plate-tectonics-seafloor-cartography.

207 told her to redo all the work
Tharp, "Connect the Dots."

207 ocean depth data to which she had access
Tharp, "Connect the Dots."

208 "seeing is believing"
Tharp, "Connect the Dots."

208 abyssal plains to the rugged mountains
Tharp, "Connect the Dots."

208 if all the water were drained away
Tharp, "Connect the Dots."

208 as bottomless as the crew's appetite
Tharp, "Connect the Dots."

208 "like the cartographers of old"
Tharp, "Connect the Dots."

209 safe locations for new telephone cables
Jeff Hect, "Laser," *Encyclopaedia Britannica,* 18 May 2023, https://www.britannica.com/technology/laser.

209 connecting users across continents
James Griffiths, "The Global Internet Is Powered by Vast Undersea Cables. But They're Vulnerable," CNN, 26 July 2019, https://edition.cnn.com/2019/07/25/asia/internet-undersea-cables-intl-hnk/index.html.

209 major internet disruptions on the US East Coast
University of Southern California, "Internet Outages in the US Doubled during Hurricane Sandy," ScienceDaily, 18 December 2012, www.sciencedaily.com/releases/2012/12/121218133152.htm.

209 had destroyed some of their transatlantic cables
Tharp, "Connect the Dots."

210 compares to the Copernican revolution
Tharp, "Connect the Dots."

210 or even the 1980s in some countries
Naomi Oreskes, "History and Memory," in *Plate Tectonics: An Insider's History of the Modern Theory of the Earth,* ed. Naomi Oreskes (Boulder, CO: Westview Press, 2003), xi–xxiv.

211 we haven't yet been able to get very deep
JOIDES Resolution, International Ocean Discovery program, n.d., accessed 2 June 2023, https://joidesresolution.org/.

211 inside the Earth, where these rocks formed
Hank Green, "What's Actually Inside the Earth's Core? | Journey to the Center of the Earth," YouTube video, posted by SciShow, 1 March 2020, https://www.youtube.com/watch?v=tquABLc3Hhs.

211 billions of people live on a ticking bomb
"IBM-backed Grillo open sources earthquake early-warning system through The Linux Foundation," IBM Developer blog, 10 August 2020, https://developer.ibm.com/blogs/ibm-backed-grillo-open-sources-earthquake-early-warning-system-openeew/.

211 Chinese scientist named Chang Heng
"What Was the First Instrument That Actually Recorded an Earthquake?" FAQs, US Geological Survey, n.d., accessed 2 June 2023, https://www.usgs.gov/faqs/what-was-first-instrument-actually-recorded-earthquake#:~:text=The.

212 called seismometers or seismographs
"Seismometers, Seismographs, Seismograms—What's the Difference? How Do They Work?" FAQs, US Geological Survey, n.d., accessed 2 June 2023, https://www.usgs.gov/faqs/seismometers-seismographs-seismograms-whats-difference-how-do-they-work.

212 wiggles on the paper
"How a Seismometer Works," SEIS InSight, last updated 7 November 2016, https://www.seis-insight.eu/en/public-2/planetary-seismology/how-a-seismometer-works.

213 interested in Himalayan earthquakes
"Richard Dixon Oldham," in course materials, Eliza Richardson, Earth 520: Plate Tectonics and People, Spring 2021, Department of Geosciences, Penn State University, accessed 2 June 2023, https://www.e-education.psu.edu/earth520/node/1782.

213 surface waves, P waves and S waves
Christopher DeCou, "This Week in the History of Science: Richard Dixon Oldham, the Discoverer of the Earth's Core and Earthquake Waves," Medium, 1 August 2018, https://medium.com/@ccdecou/this-week-in-the-history-of-science-1ac7e1d3d4e.

213 like a Slinky moving up and down
Justin Leung, "Mathsquake—the Maths of Earthquakes," Tom Rocks Maths, 27 February 2010, https://tomrocksmaths.com/2021/10/11/mathsquake-the-maths-of-earthquakes/.

214 is captured by Snell's law
Jeremy Norman, "Ibn Saul Discovers the Law of Refraction," Jeremy Norman's History of Information, 9 May 2023, https://www.historyofinformation.com/detail.php?id=2048.

215 seismic stations all over the world
"Seismic Wave," *Encyclopaedia Britannica,* last updated 30 May 2023, https://www.britannica.com/science/seismic-wave.

215 would suddenly stop, as in Figure 8.4
Karla Panchuk, "Imaging Earth's Interior," in *Physical Geology,* 1st USask ed. (Saskatoon: First University of Saskatchewan, 2019), https://openpress.usask.ca/physicalgeology/.

216 Nobel Prize laureate Niels Bohr
Marie D. Eriksen, "Inge Lehmann, part 1: Inge Lehmann and the Interior of the Earth," Great Danish Researchers, Niels Bohr Institute, University of Copenhagen, 1 September 2014, https://nbi.ku.dk/english/www/inge/lehmann/indledning/.

217 adamant about treating girls and boys equally
"Hanna Adler, Margrethe Bohr, Hans Bohr, Ernest Bohr, Aage Bohr, Rigmor Adler and Erik Bohr," photograph, 1932, Billeder (Photograph) B935, Niels Bohr Archive, ArkivDK, accessed 2 June 2023, https://arkiv.dk/vis/6013480.

217 this was not the popular opinion
Eriksen, "Inge Lehmann and the Interior of the Earth."

217 "amongst boys and young men at home"
Bruce A. Bolt, "Inge Lehmann," *Biographical Memoirs of Fellows of the Royal Society* 43 (1997): 287–301, https://doi.org/10.1098/rsbm.1997.0016.

218 posed a major challenge
Marie D. Eriksen, "Inge Lehmann, part 2: Geodetic Institute," Great Danish Researchers, Niels Bohr Institute, University of Copenhagen, 1 October 2014, https://nbi.ku.dk/english/www/inge/lehmann/andet-kap/.

218 would have mostly consisted of oceans
Antipodes map, accessed 2 June 2023, https://www.antipodesmap.com/.

219 a paper entitled simply "P′"
From Inge Lehmann's 1936 paper "P′," *Publications du Bureau Central Séismologique International,* series A, Travaux Scientifique, vol. 14 (1936): 87–115.

219 and then the mantle again
Lehmann, "P′."

219 international exchange of data became all but impossible
Marie D. Eriksen, "Inge Lehmann, part 4: The Last Years at the Institute," Great Danish Researchers, Niels Bohr Institute, University of Copenhagen, 3 February 2015, https://nbi.ku.dk/english/www/inge/lehmann/the-last-years/.

219 detect underground nuclear explosions
Bolt, "Inge Lehmann."

219 type of surface seismic wave
Eriksen, "Inge Lehmann, Last Years."

220 new theories of the inner parts of the Earth
Bolt, "Inge Lehmann."

220 the seismic vibrations of elephants
Paula Koelemeijer, "Outreach," Paula Koelemeijer, Royal Society University Research Fellow, University of Oxford, accessed 4 February 2024, https://www.earth.ox.ac.uk/~univ4152/outreach.html.

221 graphō *("to write, to describe")*
"Tomography," Academic Dictionaries and Encyclopedias, accessed 2 June 2023, https://etymology.en-academic.com/35148/tomography.

221 two-dimensional cross-sections
Sid Perkins, "Seismic Tomography Uses Earthquake Waves to Probe the Inner Earth," *Proceedings of the National Academy of Sciences* 116, no. 33 (2019): 16159–16161, https://doi.org/10.1073/pnas.1909777116.

221 an average over these variations
Philip Kearey, Keith A. Klepeis, and Frederick J. Vine, *Global Tectonics,* 3rd ed. (Chichester: John Wiley, 2009).

222 corresponding to thousands of earthquakes
Jeroen Tromp, "Imaging the Earth's Interior with the Summit Supercomputer," NVIDIA Developer News Center, YouTube video, 9 November 2018, https://www.youtube.com/watch?v=1wOxSfESmsY.

223 outside of tectonic plate boundaries
"Hot Spots," National Geographic Society, Education, n.d., accessed 2 June 2023, https://education.nationalgeographic.org/resource/hot-spots/.

223 fueled by mantle plumes
"Deep Earth Research," Deep Earth Explorers, Sanne Cottaar / Deep Earth Seismology research group, Department of Earth Sciences, University of Cambridge, n.d., accessed 2 June 2023, https://earth-deepearth.esc.cam.ac.uk/deep-earth-explorers/deep-earth-research-3/.

223 precise paths of seismic waves
Inge Lehmann, "Seismology in the Days of Old," *Eos, Transactions American Geophysical Union* 68, no. 3 (1987): 33–35, https://doi.org/10.1029/eo068i003p00033-02.

224 and its epicenter are approximately the same
William Lowrie, *Fundamentals of Geophysics,* 2nd ed. (Cambridge: Cambridge University Press, 2007), 150.

224 how this process got its name: trilateration
Leung, "Mathsquake."

224 twenty-four satellites orbiting our planet
"Space Segment," GPS.gov, accessed 2 June 2023, https://www.gps.gov/systems/gps/space/.

225 Latin planitia—*flat surface*
"Insight's Landing Site: Elysium Planitia," NASA / JPL-Caltech, 18 March 2019, https://solarsystem.nasa.gov/resources/861/insights-landing-site-elysium-planitia/.

225 a whole seismological network
"Instruments: Seismometer," Mars InSight Mission, NASA, n.d., accessed 2 June 2023, https://mars.nasa.gov/insight/spacecraft/instruments/seis/.

225 it had recorded 1,319 marsquakes
Karen Fox, Alana Johnson, and Andrew Good, "NASA Retires Insight Mars Lander Mission after Years of Science," press release, NASA, 21 December 2022, https://www.nasa.gov/press-release/nasa-retires-insight-mars-lander-mission-after-years-of-science.

225 movements of magma
Jonathan O'Callaghan, "Insight Lander Makes Best-Yet Maps of Martian Depths," *Scientific American,* 22 July 2021, https://www.scientificamerican.com/article/insight-lander-makes-best-yet-maps-of-martian-depths/.

225 seismometers thousands of miles away
"Journey to the Center of Mars with the Lander Team," InSight Live Q&A, NASA Jet Propulsion Laboratory, 23 July 2021, YouTube video, https://www.youtube.com/watch?v=kca3Y8XUK1c&t=335s.

226 reflected off different layers of Mars
"Insight Mission: Mars Unveiled," EurekAlert! 22 July 2021, https://www.eurekalert.org/news-releases/876661.

226 which ones best matched the observations
"Insight Mission: Mars Unveiled."

226 the core is much larger than expected
O'Callaghan, "InSight Lander Makes Best-Yet Maps."

226 understand the origin of rocky planets
"Journey to the Center of Mars with the Lander Team."

227 InSight *took just two years to measure Mars' core*
Brandon Specktor, "Scientists Mapped the Mysterious Interior of Mars for the First Time Ever," LiveScience, 22 July 2021, https://www.livescience.com/mars-interior-first-map.html.

227 260 billion cubic miles of the Earth's interior
Tariq Khokhar, "7 Things You May Not Know about Water," World Bank Data Blog, 6 September 2013, https://blogs.worldbank.org/opendata/7-things-you-may-not-know-about-water.

227 undersea volcanoes that remain to be mapped
Paul Voosen, "'It's Just Mind Boggling.' More Than 19,000 Undersea Volcanoes Discovered," *Science,* 19 April 2023, doi: 10.1126/science.adi3418.

227 a distinct layer within the Earth's inner core
Thanh-Son Phạm and Hrvoje Tkalčić, "Up-to-Fivefold Reverberating Waves through the Earth's Center: Distinctly Anisotropic Innermost Inner Core," *Nature Communications* 14 (2023), art. 754, https://doi.org/10.1038/s41467-023-36074-2.

Conclusion

228 with a simple click on an app
"Seeing the Road Ahead," Waymo, "About," accessed 7 June 2023, https://waymo.com/company/.

229 lidars, radars, cameras, and other sensors
"How Autonomous Vehicles Work," Waymo, n.d., accessed 7 June 2023, https://ltad.com/about/how-autonomous-vehicles-work.html#:~:text=LiDAR%20System,and%20return%20to%20the%20vehicle.

229 stop signs, and lane markers
"Waymo Driver," Waymo, n.d., accessed 7 June 2023, https://waymo.com/waymo-driver/.

229 manually added by engineers

Saket Sonekar, "Are HD Maps a Bottleneck for Self-Driving Cars?" LinkedIn, 19 September 2021, https://www.linkedin.com/pulse/hd-maps-bottleneck-self-driving-cars-saket-sonekar/.

229 to alter the map accurately

Divya Agarwal, "How SLAM Works for Self-Driving Cars: A Brief but Detailed Overview," AutoVision News, 5 June 2020, https://www.autovision-news.com/adas/how-slam-works/.

FURTHER READING

This book wouldn't exist without the hundreds of wonderful books, journal articles, websites, podcasts, and videos I've consulted during my research. If you're interested in learning more about some topics, here you can find a selection of the most relevant or comprehensive sources.

General

Harley, J. B., David Woodward, et al., eds. *History of Cartography,* vols. 1–4 and 6. Chicago: University of Chicago Press, 1987–2021.

1. Curved

Christopher Columbus

Morison, Samuel Eliot. *Admiral of the Ocean Sea: A Life of Christopher Columbus.* Boston: Little, Brown, 1942.

Rickey, V. Frederick. "How Columbus Encountered America." *Mathematics Magazine* 65, no. 4 (October 1992): 219–225, https://doi.org/10.1080/0025570x.1992.11996024.

Triangulation

Ferreiro, Larrie D. *Measure of the Earth: The Enlightenment Expedition That Reshaped Our World.* New York: Basic Books, 2011.

Haasbroek, N. D. *Gemma Frisius, Tycho Brahe and Snellius and Their Triangulations.* Delft: Rijkscommissie voor Geodesie Delft, 1968.

Gauss and Spherical Geometry

Bühler, Walter K. *Gauss: A Biographical Study.* New York: Springer-Verlag, 1981.

Dunnington, G. Waldo, and Jeremy Gray. *Gauss: Titan of Science.* Washington, DC: Mathematical Association of America, 2004.

Spivak, Michael. *A Comprehensive Introduction to Differential Geometry,* vol. 2, 3rd ed. Houston: Publish or Perish, 1999.

2. Flat

Map Projections

Klinghoffer, Arthur Jay. *The Power of Projections: How Maps Reflect Global Politics and History.* Westport, CT: Praeger, 2006.

Snyder, John P. *Map Projections: A Working Manual.* U.S. Geological Survey Professional Paper 1395. Washington, DC: U.S. Government Printing Office, 1987, https://pubs.usgs.gov/pp/1395/report.pdf.

Tobler, Waldo. "Qibla, and Related, Map Projections." *Cartography and Geographic Information Science* 29, no. 1 (2002): 17–23, https://doi.org/10.1559/152304002782064574.

Gerardus Mercator and His Projection

Monmonier, Mark S. *Rhumb Lines and Map Wars: A Social History of the Mercator Projection.* Chicago: University of Chicago Press, 2004.

Rickey, V. Frederick, and Philip M. Tuchinsky. "An Application of Geography to Mathematics: History of the Integral of the Secant." *Mathematics Magazine* 53, no. 3 (1980): 162–166, https://doi.org/10.1080/0025570x.1980.11976846.

Taylor, Andrew. *The World of Gerard Mercator: The Mapmaker Who Revolutionised Geography.* London: Harper Perennial, 2005.

3. Scaled

Lewis Fry Richardson

Hunt, J. C. R. "Lewis Fry Richardson and His Contributions to Mathematics, Meteorology, and Models of Conflict." *Annual Review of Fluid Mechanics* 30, no. 1 (1998): xiii–xxxvi, https://doi.org/10.1146/annurev.fluid.30.1.0.

West, Geoffrey B. *Scale: The Universal Laws of Growth, Innovation, Sustainability, and the Pace of Life in Organisms, Cities, Economies, and Companies.* New York: Penguin Press, 2017.

Benoit Mandelbrot and Fractals

Gleick, James. *Chaos: Making a New Science.* New York: Viking, 1987.

Lesmoir-Gordon, Nigel, ed. *The Colours of Infinity: The Beauty and Power of Fractals,* 2nd ed. London: Springer, 2010.

Mandelbrot, Benoît. "Drawing; The Ability to Think in Pictures and Its Continuing Influence." Web of Stories videos, 24 January 2008, https://www.webofstories.com/play/benoit.mandelbrot/8.

Coastline Paradox

Stoa, Ryan B. "The Coastline Paradox." *Rutgers University Law Review* 72, no. 2 (Winter 2019): 351–400.

4. Distanced

Underground Map

Guo, Zhan. "Mind the Map! The Impact of Transit Maps on Path Choice in Public Transit." *Transportation Research Part A: Policy and Practice* 45, no. 7 (2011): 625–639, https://doi.org/10.1016/j.tra.2011.04.001.

Kent, Alexander J. "When Topology Trumped Topography: Celebrating 90 Years of Beck's Underground Map." *Cartographic Journal* 58, no. 1 (2021): 1–12, https://doi.org/10.1080/00087041.2021.1953765.

Hamming Distance

Thompson, Thomas M. *From Error Correcting Codes through Sphere Packings to Simple Groups.* Washington, DC: Mathematical Association of America, 1983.

Neuroscience of Navigation

Bond, Michael. *From Here to There: The Art and Science of Finding and Losing Our Way.* Cambridge, MA: The Belknap Press of Harvard University Press, 2020.

Epstein, Russell A., et al. "The Cognitive Map in Humans: Spatial Navigation and Beyond." *Nature Neuroscience* 20, no. 11 (2017): 1504–1513, https://doi.org/10.1038/nn.4656.

Moser, May-Britt, and Noa Segev. "How Do We Find Our Way? Grid Cells in the Brain." Frontiers for Young Minds, 7 September 2021, https://kids.frontiersin.org/articles/10.3389/frym.2021.678725.

Warren, William H. "Non-Euclidean Navigation." *Journal of Experimental Biology* 222, suppl. 1 (2019), https://doi.org/10.1242/jeb.187971.

5. Connected

Seven Bridges of Königsberg

Biggs, Norman, Edward K. Lloyd, and Robin J. Wilson. *Graph Theory 1736–1936.* Oxford: Clarendon Press, 1986.

Dunham, William. *The Genius of Euler: Reflections on His Life and Work.* Washington, DC: Mathematical Association of America, 2007.

Hopkins, Brian, and Robin J. Wilson. "The Truth about Konigsberg." *College Mathematics Journal* 35, no. 3 (2004): 198–207, https://doi.org/10.2307/4146895.

Traveling Salesman Problem

Cook, William J. *In Pursuit of the Traveling Salesman: Mathematics at the Limits of Computation.* Princeton: Princeton University Press, 2011.

Four Color Theorem

Appel, Kenneth, and Wolfgang Haken. "The Solution of the Four-Color-Map Problem." *Scientific American* 237, no. 4 (1977): 108–121, https://doi.org/10.1038/scientificamerican1077-108.

Wilson, Robin. *Four Colors Suffice,* rev. ed. Princeton: Princeton University Press, 2013.

6. Divided

Gerrymandering

Cho, Wendy K. Tam, and Yan Y. Liu. "Toward a Talismanic Redistricting Tool: A Computational Method for Identifying Extreme Redistricting Plans." *Election Law Journal: Rules, Politics, and Policy* 15, no. 4 (2016): 351–366, https://doi.org/10.1089/elj.2016.0384.

Engstrom, Erik J. *Partisan Gerrymandering and the Construction of American Democracy.* Ann Arbor: University of Michigan Press, 2013.

Orcutt, Mike. "How Math Has Changed the Shape of Gerrymandering." *Quanta Magazine,* 1 June 2023, https://www.quantamagazine.org/how-math-has-changed-the-shape-of-gerrymandering-20230601/.

Seabrook, Nicholas R. *One Person, One Vote: A Surprising History of Gerrymandering in America.* New York: Pantheon Books, 2022.

Stephanopoulos, Nicholas O., and Eric M. McGhee. "Partisan Gerrymandering and the Efficiency Gap." *University of Chicago Law Review* 82, no. 2 (2015): 831–900.

School Segregation

Chang, Alvin. "School Segregation Didn't Go Away. It Just Evolved." Vox, 27 July 2017, https://www.vox.com/policy-and-politics/2017/7/27/16004084/school-segregation-evolution.

EdBuild. https://edbuild.org/, accessed 1 June 2023.

Monarrez, Tomás, and Carina Chien. "Dividing Lines: Racially Unequal School Boundaries in US Public School Systems," report, Urban Institute, 1 September 2021, https://www.urban.org/research/publication/dividing-lines-racially-unequal-school-boundaries-us-public-school-systems.

Monarrez, Tomas, et al. "How Much Does Your School Contribute to Segregation?" Urban Institute, 8 July 2020, https://apps.urban.org/features/school-segregation-index/.

7. Found

Epidemiology

Field, Kenneth. "Something in the Water: The Mythology of Snow's Map of Cholera." ArcGIS blog, Environmental Systems Research Institute (ESRI), Redlands, CA, 3 December 2020, https://www.esri.com/arcgis-blog/products/arcgis-pro/mapping/something-in-the-water-the-mythology-of-snows-map-of-cholera/.

Johnson, Steven. *The Ghost Map: The Story of London's Most Terrifying Epidemic—and How It Changed Science, Cities, and the Modern World.* New York: Riverhead Books, 2006.

Pastore y Piontti, Ana, Nicola Perra, Luca Rossi, Nicole Samay, and Alessandro Vespignani. *Charting the Next Pandemic: Modeling Infectious Disease Spreading in the Data Science Age.* Cham, Switzerland: Springer, 2019.

Snow, John. *On the Mode of Communication of Cholera,* 2nd ed. London: John Churchill, 1855.

Geographic Profiling

Chainey, Spencer, and Lisa Tompson, eds. *Crime Mapping Case Studies: Practice and Research.* Chichester: John Wiley and Sons, 2008.

Rossmo, D. Kim. *Geographic Profiling.* Boca Raton, FL: CRC Press, 2000.

Rossmo, D. Kim. "Recent Developments in Geographic Profiling." *Policing: A Journal of Policy and Practice* 6, no. 2 (9 June 2012): 144–150, https://doi.org/10.1093/police/par055.

Bayesian Search

Stone, Lawrence D., Colleen M. Keller, Thomas M. Kratzke, and Johan P. Strumpfer. "Search for the Wreckage of Air France Flight AF 447." *Statistical Science* 29, no. 1 (February 2014): 69–80, https://doi.org/10.1214/13-sts420.

8. Deep

Mapping the Ocean Floor

"Nautilus Live." https://nautiluslive.org/, accessed 4 October 2023.

Tharp, Marie. "Connect the Dots: Mapping the Seafloor and Discovering the Mid-Ocean Ridge." In *Lamont-Doherty Earth Observatory: Twelve Perspectives on the First Fifty Years,* ed. Laurence Lippsett. Palisades, NY: Lamont-Doherty Earth Observatory of Columbia University, 1999.

Mapping the Earth's Interior

Eriksen, Marie D. "Inge Lehmann." 6 parts. Great Danish Researchers, Niels Bohr Institute, University of Copenhagen, 11 September 2014–9 February 2015, https://nbi.ku.dk/english/www/inge/lehmann/.

Lehmann, Inge. "Seismology in the Days of Old." *Eos, Transactions American Geophysical Union* 68, no. 3 (1987): 33–35, https://doi.org/10.1029/eo068i003p00033-02.

Lowrie, William. *Fundamentals of Geophysics,* 2nd ed. Cambridge: Cambridge University Press, 2007.

Oreskes, Naomi. "History and Memory." In *Plate Tectonics: An Insider's History of the Modern Theory of the Earth,* ed. Naomi Oreskes. Boulder, CO: Westview Press, 2003.

Mapping Mars's Interior

Cottaar, Sanne, and Paula Koelemeijer. "The Interior of Mars Revealed." *Science* 373, no. 6553 (2021): 388–389, https://doi.org/10.1126/science.abj8914.

ACKNOWLEDGMENTS

Of all the sections of this book, none have I procrastinated on longer than the acknowledgments. How do I express my gratitude to all the people who have enabled me to fulfill my childhood dream of publishing a book? How can I summarize in just a few sentences how thankful I am for having you in my life? Here's my imperfect attempt to recognize those who have helped me to turn a vague idea into a real book.

I knew Andrea Henry was the perfect editor for me the first time we spoke. Thanks to her for her constructive feedback and for calling me out whenever I made the math more complicated than necessary. Her edits have made my book a thousand times better. Thanks also to Rachel Field, whose involvement I appreciated from the beginning, especially her detailed feedback on the math-heavy parts. I'm grateful to both of my editors for making such a great editorial team. Big thanks also to everyone at Picador and Harvard University Press who has worked behind the scenes to bring my book into the world. Thank you for taking care of the illustrations, design, copyedits, rights, publicity, marketing, and everything else I had no idea was even needed to publish and sell a book.

Thank you to everyone who kindly agreed to share their expertise with me. Every single interview was extremely helpful, no matter how much of it ended up on the book's pages. Sam Arlin, Dorothea Blostein, Wendy K. Tam Cho, Sanne Cottaar, Matthew Edney, Larrie Ferreiro, Tom Koch, Paula Koelemeijer, Tomás Monarrez, Kim Rossmo, Dan Saunders, Derek Sowers, Hugo Spiers, Colleen Sterling, and Ryan Stoa—this book wouldn't be the same without your voices.

I would never have attempted to write a book if it weren't for Neuwrite London. I still remember our first meeting at the pub, when I was too shy to speak without being asked, feeling starstruck by all those fabulous, published authors. Thank you for sharing your amazing writing, for reading my half-baked proposals and chapters, and for supporting me through the ups and downs of the publishing process. Special thanks to Roma Agrawal for introducing me to Rukhsana Yasmin, the literary agent who first took me under her wing. The world of publishing can be a harsh environment for new authors, and I wouldn't have been able to navigate it without my two excellent agents. Rukhsana, thank you for believing in my idea and championing my book with publishers. And thank you to Amandeep Singh for taking over in these crucial final months and helping me get *Mapmatics* across the finish line.

Writing a book while working full time is hard—but working with a writer who keeps taking days off to meet editorial deadlines and turning up to meetings half asleep after working on their book into the early hours must be even harder. Thank you to my team at Brilliant for creating a safe and understanding environment and for cheering me on as I worked on *Mapmatics*. A special thanks to Michelle McSweeney for being such a kind and accommodating manager and for adjusting my workload to give me a chance to finish the book.

To all my friends, thank you for checking on me despite my canceling so many outings. Marlena, thank you for reading parts of the book (and for having a wedding that inspired the section about seating plans and graphs). Czarek, I appreciate your feedback on the particularly difficult sections and your encouragement to keep the jokes in. Zack, your input on the jacket design was invaluable—I know, I know, I still owe you chocolates!

I'm grateful to my family for instilling in the little me a love of books and a curiosity about the world. Mum, Dad, Grandma Ala, Grandma Halina, and Grandpa Zygmunt: thank you for encouraging my passions and celebrating my achievements. And I apologize for all the Sunday lunches I've skipped to finish up a chapter! Huge thanks to my lovely rescue dog Koala for distracting me with her cute ears and reminding me to take cuddle breaks. No, you can't eat this book, but I'll buy you a special treat, I promise!

There's one person who contributed to this book more than anyone else. Marco, thank you for supporting me throughout the whole process, for believing in me and the book, for reading its many versions, for eagerly discussing even the most obscure math, for accompanying my writing sessions, for comforting me when I just couldn't get that paragraph right, for bringing me coffee whenever I needed it (or looked like I could do with some), and for taking care of Koala when I struggled to meet the deadlines. I would have given up a thousand times without your support. I'm so glad you dragged me to that pizzeria on the other side of London all those years ago!

I'm a reader first and only then a writer. Thank you to my fellow writers, fiction and nonfiction, for inspiration and the joy your books bring me. It's reassuring to know that when I need an escape from work—or life—I can find a story that will transport me to a different reality. Reading has kept me sane while I worked on *Mapmatics,* which I'm immensely grateful for. Books make the world a better place—please keep writing!

ILLUSTRATION CREDITS

Figure 1.9 "Geodesics on the Earth: Demonstrating How Geodesics Appear to Curve on a 2D Representation of 3D Space," *Academo,* https://academo.org/demos/geodesics, accessed 10 January 2024.

Figure 2.4 V. Frederick Rickey and Philip M. Tuchinsky, "An Application of Geography to Mathematics: History of the Integral of the Secant," *Mathematics Magazine 53,* no. 3 (1980): 162–166, 163.

Figure 2.6 Stefan Kühn, "Mercator Projection Map with Tissot's Indicatrices," https://commons.wikimedia.org/wiki/File:Tissot_mercator.png, accessed 25 August 2023, CC BY-SA 3.0.

Figure 2.7 Daniel R. Strebe, "Part of the World on Craig Retroazimuthal Projection. 15° Graticule, Central Meridian 39°49′E, center latitude 21°25′N (Mecca)," https://en.wikipedia.org/wiki/File:Craig_projection_SW.jpg, accessed 25 August 2023, CC BY-SA 3.0.

Figure 3.2 L. F. Richardson, "The Problem of Contiguity: An Appendix to Statistics of Deadly Quarrels," *General System Yearbook* 6 (1961): 139–187, 169.

Figure 4.1 "Underground Electric Railways of London Co. Ltd" © TfL, from the London Transport Museum collection, "Pocket Underground map issued by UERL" (1908), https://www.ltmuseum.co.uk/collections/collections-online/maps/item/2002-264, accessed 20 November 2023.

Figure 4.2 Francis Nicholson, "Map of the Several Nations of Indians to the Northwest of South Carolina," or the Catawba Deerskin Map [S.I.: s.n] (1724), retrieved from the Library of Congress, https://www.loc.gov/item/2005625337/, accessed 1 June 2023.

Figure 4.3 Dan Saunders, "Crow flies, why do people still use the circle-based approach?," Basemap (2017), retrieved from the Wayback Archive, https://web.archive.org/web/20171008040826/http://www.basemap.co.uk/crow-flies/, accessed 13 February 2024.

Figure 5.1 Matthäus Merian-Erben/Alamy, "Map of Königsberg" (1652), https://www.alamy.com/image-koenigsberg-map-by-merian-erben-1652-image356483670.html.

Figure 5.2 Leonhard Euler, "Solutio Problematis Ad Geometriam Situs Pertinentis," *Commentarii Academiae Scientiarum Imperialis Petropolitanae* 8 (1736): 128–140, illus. foll. 158.

Figure 5.6 Santacloud, "Present State of the Seven Bridges of Königsberg," adapted from rendering by GIScience Research Group @ Heidelberg University, https://commons.wikimedia.org/wiki/File:Present_state_of_the_Seven_Bridges_of_K%C3%B6nigsberg.png, CC BY-SA 2.5.

Figure 5.12 Adapted from Augustus de Morgan, letter to Rowan Wilson Hamilton (1852). TCD MS 1493, 668, Trinity College Dublin Library, Manuscripts Department.

Figure 6.1 League of Women Voters of Asheville-Buncombe County, "Gerrymander 5K route" (2017), https://www.npr.org/sections/thetwo-way/2017/10/30/560909678/not-so-fun-run-gerrymander-5k-joggers-trace-asheville-s-electoral-districts, accessed 14 February 2024.

Figure 7.2 John Snow, *On the Mode of Communication of Communication of Cholera,* 2nd ed. (London: John Churchill, 1855), illus. foll. 44. Digitally enhanced version courtesy of the UCLA Department of Epidemiology.

Figure 8.4 Steven Earle, "Understanding Earth Through Seismology," *Physical Geology,* 2nd ed. (Victoria, BC: BCampus, 2019), retrieved from https://opentextbc.ca/physicalgeology2ed/chapter/9-1-understanding-earth-through-seismology/, CC BY-SA 4.0.

INDEX

The letter *f* following a page number denotes a figure; *n* denotes a footnote; *t* denotes a table.

abstraction, mathematics and, 93
Adler, Hanna, 216–217
Admiral of the Ocean Sea (Morison), 11
aesthetics, Rand McNally world map and emphasis on, 53–56
air flights, great circles and, 27–30, 28f
Air France Flight 447, search for wreckage of, 191, 192–193, 195–198
Alaska: aviation and great circles and, 29–30; borders of Alaskan Panhandle and, 83–86
algorithms, 124; ant colonization optimization, 129; attendance zone optimization, 171; evolutionary, 157; flip MCMC, 158; geographic profiling, 189; MCMC, 157–158; Meals on Wheels, 126–128, 127f; nearest neighbor, 124–125, 125f; ORION, 130–131; used to detect gerrymandering, 155–159
analysis, 142n
animation, fractals and, 81–83
ant colonies, biomimetics and, 128–129
ant colonization optimization algorithm, 129
Apian, Peter, 11
Appel, Kenneth, 140–141, 142
"Application of Geography to Mathematics, An" (Rickey and Tuchinsky), 41
area: distortion of on Mercator projection, 2, 44–47, 45f; mapmaking choices and preservation of, 33
Aristotle, 5–6
artificial intelligence, proofs done by, 143
Asheville (North Carolina), Gerrymander 5K, 144–145, 145f, 147–148
attendance zones, school segregation and, 166, 169–171, 172
autonomous cars, maps and, 228–230
azimuthal projections, 35, 35f; politics and, 61–62; retroazimuthal projections, 58–59, 59f, 60f

backscatter, 204
balloon analogy for Mercator projection, 42
Baltimore phenomenon, 67
banana, curvature of, 20–21, 21f
Bartholdi, John H., III, 126–127
baseline: Geodesic Mission to the Equator and, 17; national waters and, 86–87; in triangulation, 12, 12f
Bayes, Thomas, 191
Bayesian search, 190–198
Bayes's theorem, 191
Bears Ears site, 162–163
Beck, Henry, Tube map and, 91–92, 93, 94, 116, 230
Behrmann, Walter, 48
Bell Telephone Laboratories (BTL), 209
Berann, Heinrich, 210
Bessel, Friedrich Wilhelm, 19
bias, mapmaker, 3
Biden, Joe, 163
binary code word, 107
biomimetics, 128–129
Blostein, Dorothea, 141, 142
Bohr, Niels, 216–217
Bolivar, Simón, 18

Bond, Henry, 43
Bonin, Pierre-Cédric, 197–198
borders: Alaskan, 83–86; Mandelbrot and, 73–74; Richardson and, 69–71; western United States, 67–68. *See also* coastline paradox
Boston Public Schools, Gall-Peters projection use in, 32, 49–50
Brantingham, Patricia, 186, 188
Brantingham, Paul, 186, 188
Britain, measuring coastline of, 73–74. *See also* England
Broad Street pump, transmission of cholera and, 177–180, 178f, 181
Brown, Oliver, 165
Brown v. Board of Education, 165, 167
Budd, William, 180–181
Bureau d'Enquêtes et d'Analyses (BEA), 196–197
burglaries, using geographic profiling to solve, 189–190
Büttner, J. G., 18–19

calculus, 43, 44, 75n, 142n
Canada, Alaskan Panhandle border and, 84–86
cannon-shot rule, 86
Cantor set, 72
Carpenter, Loren, 81–82
Carroll, Lewis, 65
cartographic generalization, 66
cartography, defined, 65
Catawba Deerskin Map, 96–97, 97f
Catmull, Ed, 82
Cayley, Arthur, 136–137
Center for Geospatial Intelligence and Investigation, 185
Certaine Errors in Navigation (Wright), 40–41
"chair burglar," 189–190
Chang Heng, 211–212
Cheffins, Charles, 177n
Cho, Wendy K. Tam, 154–157, 158–159
cholera, maps and theories regarding transmission of, 173–181
Christian Aid, Peters's map and, 49
circles, measuring, 76, 77
circles of latitude, 2f
Clausewitz, Carl von, 15
Clomedes, 6
closeness, measuring, 101–104
coastline paradox, 70–71, 73–74; Alaskan Panhandle borders and, 84–86; legal implications of, 86–89; Steinhaus and, 80–81
cognitive graph, 111–113
cognitive map, 108–113
Cold War: flying through Alaska and, 29; power of projections used during, 61–62
Columbus, Christopher, 5, 37; plotting route to East Asia, 6, 9–11
Commissioners' Plan of 1811, 99–100
compartmental models, 182–183
computers: correcting code, 105–108; detecting gerrymandering using, 154–159, 156f; graphs and, 143; validity of calculations on, 143
conflict, geography and, 69
conformal projections, 33
conic projections, 34–35, 35f
connections: ant colony and, 128–129; delivery route scheduling, 121–124, 126–131; Euler and, 114–120; five-color theorem, 137–140; four-color conjecture, 133–137, 135f; four-color theorem, 136–137, 139–143; graph theory, 120–121; Meals on Wheels heuristic, 126–128; nearest neighbor algorithm, 124–125, 125f; seating plan graphs, 131–133, 132f; simplifying maps, 116–120, 117f; topology and, 93
contiguous zone, 87, 88f
continental drift: evidence for, 207–211; Tharp's discovery and, 207–208, 210–211
Cook, Rob, 82
Core-Mantle Boundary, 220
Cosmographia (Apian), 11, 13
Cottaar, Sanne, 223
Cousteau, Jacques, 209–210
COVID-19 epidemic, mapping, 182
cracking, 146–147
Craig, James Ireland, 58
Craig retroazimuthal projection, 58, 59f
crime investigation, geographic profiling and, 185–190
curvature, 20–22; Gaussian, 22–25
curves, measuring, 75–78
cylinder, curvature of, 20–22, 22f
cylindrical projections, 34, 35f; Mercator projection and, 37, 38f; Peter's map and, 48–49

Darwin, Charles, 18
da Vinci, Leonardo, 128
Dee, John, 46–47
De Morgan, Augustus, 134–136, 135f
Descartes, René, on shape of the Earth, 14
De Witt, Simeon, 99–100
differential geometry, 1–2
dimension: defined, 78–79; fractional, 79–81

Disquisitiones generals circa superficies curvas (Gauss), 20
distance: calculating on Mercator map, 42–44, 43f; cognitive map, 108–113; defined, 104; driving, 101; Euclidean, 101–102, 103–104; finding fastest routes, 94–98; Hamming, 107–108; Legible London, 98–99; map-making choices and preservation of, 33–34; metric space, 104–108; navigating New York and, 99–104; sense of, 110–111; straight, 101; topology, 92–94; Tube map, 90–92, 91f, 93, 94–95; walking, 98–99, 100f, 101
distance functions, 101–104; Manhattan metric, 103–104; properties of, 102
distortions, map, 32; educating students about, 49–51; Mercator projection and, 2, 44–47, 45f. *See also* scale
Dodgson, Charles Lutwidge, 65n
Doob, Joseph L., 142
Dorigo, Marco, 129
Dostrovsky, Jonathan, 109
drómos, 37
Duchin, Moon, 158

Earmuff District (Illinois), 159–161, 160f
Earth: ancient Greeks on spherical shape of, 5–7; circumference of, 6–8, 10; composition of core, 215–216, 218–220; composition of mantle, 222–223; contemporary mapping and model of interior of, 220–223; curvature of, 25–26; debating shape of, 14–18, 30; flattening of, 34–36; mapping interior of, 211–225; radius of, 54n; using triangulation to measure, 13
earthquakes: detecting, 211–212, 212f; epicenter (*see* epicenters, seismic); mid-ocean ridges and, 209; waves of, 213–215
EdBuild, 171–172
edges, graphs and, 117–119, 121
Edney, Matthew, 19
efficiency, Tube map and, 94–96
efficiency gap, 151–154, 152t, 153t
egg, curvatures of an, 21, 21f, 22–23, 24–25
egg-shape of Earth, debate over, 14–16
Ehler, Carl Leonhard Gottlieb, 114–115
elections: effect of gerrymandering on, 145–147, 146f; using results to identify gerrymandering, 151–154, 152t, 153t
Elements (Euclid), 101
England: cholera in, 173–181; gerrymandering in, 148–149
environmental criminology, 186
epicenters, seismic, 211–212; finding, 223–224, 225f; of marsquakes, 226; seismic waves and, 216
epidemiology: compartmental models and epidemics, 182–185; Global Epidemic and Mobility model, 183–184; SIR models, 182–183; Snow and, 181
EPSG Geodetic Parameter Dataset, 52–53
equal-area projection, Peters and, 48–49
equator, 2f, 34, 44–46, 45f
equidistant projections, 33–34
Eratosthenes, 6–8, 9, 10
Eratosthenes Batavus (Snell), 13
"Error Detecting and Error Correcting Codes" (Hamming), 105
esoteric constructions, 72
Esselstyn, Blake, 147–148
Euclidean distance, 101–102, 103–104
Euclidean metric, 101–102, 103–104, 110–111
Euclid of Alexandria, 101
Euler, Leonhard, 131, 138n; graph theory and, 143; Seven Bridges of Königsberg and, 114–121, 116f, 117f, 121
Eurocentric view: Gall-Peters projection against, 49–50; Mercator projection and, 46–47, 51–52
European Cities Against School Segregation (ECASS), 172
evolutionary algorithm, 157
Ewing, Maurice, 219–220
exclusive economic zone (EEZ), 87, 88f, 204

fair district, determining, 150–151
Ferreiro, Larrie D., 14, 15
Fisheries Case, 87–88
five-color theorem, 137–139
Fleury, Pierre-Henry, 119
flip MCMC algorithm, 158
Fossett, Peggy, 194
Fossett, Steve, 193–195
Foster, Howard, 209
four-color conjecture, 133–136, 135f, 140–141
four-color theorem, 136–137, 139–143
four-color theorem proof, 141–142
fractal algorithm, for animation, 82–83
Fractal Geometry of Nature, The (Mandelbrot), 81
fractality, self-similarity and, 73
fractals, 75–78, 89; Alaskan borders and failure to notice, 83–86; animation and, 81–83; defined, 82
fractional dimensions, 78–81
Fréchet, Maurice-René, 105
Frisius, Gemma, 11–13, 36

Gall, James, 48
Gall-Peters projection, 48–50, 51
Garver, John, 55
Gauss, Carl Friedrich, 46; geodesy and, 18–20; Remarkable Theorem and, 3, 22–25
Gaussian curvature, 22–25
gender, navigational preferences and, 112–113
General Bathymetric Chart of the Oceans (GEBCO), 201
generalization: cartographic, 66; topology and, 113
Geodesic Mission to the Equator, 15–18
geodesy, 8, 14, 18–20, 30–31
geographic profiling, 185–190
geographic "true" north, 40n, 57
geometry: curvature and, 25; differential, 1–2
geometry of position, Euler on, 115
Gerry, Elbridge, 148
Gerrymander 5K (Asheville, North Carolina), 144–145, 145f, 147–148
gerrymandering: cracking and packing, 145–147; detecting using computers, 154–159, 156f; Earmuff District and, 159–161, 160f; effect of Utah's on Navajo reservation, 161–163; efficiency gap and, 151–154; history of, 148–150; identifying, 150–151; outside the United States, 149–150; prevention of, 163–164; school district, 164–172
Gilbert, Edgar N., 58
Global Epidemic and Mobility (GLEAM) model, 183–184
global position system (GPS), 11, 18, 109, 111, 131, 190–191, 195, 224, 229
globes: avoiding map distortions and teaching geography using, 50; less practical to use compared to maps, 32; Mercator making, 36; translating to two-dimensional map, 1–2, 3
gnomon, 7
Gödel, Kurt, 142
Godin, Louis, 16n, 17n
Golay, Marcel J. E., 108
Google: autonomous cars and, 228–230; translating one-dimension code to two-dimensional map, 77–78; Web Mercator, 52–53
Google Earth, 200
grapefruit-shape of Earth, debate over, 14–16, 17
graph labeling, 132–133
graphs: cognitive, 111–113; computers and, 143; edges and, 117–119, 121; Euler and invention of, 116; Markov chain Monte Carlo algorithm and, 157–158; seating plan and, 131–133, 132f; vertices and, 118–119, 118f, 121
graph theory, 120–121, 143
great circle method of determining qibla, 57–58
great circles, 27–30, 28f
Gregory, James, 44
grid cells, 109, 110
Groes, Nils, 220
Grotius, Hugo, 86
Gunter, Edward, 43
Guthrie, Francis, 133
Guthrie, Frederick, 133

Haken, Wolfgang, 140–141, 142
Hamilton, William Rowan, 121, 134–136
Hamming, Richard, 105–108
Hamming distance, 107–108
Hausdorff, Felix, 105
Heawood, Percy, 137–138, 139
Heezen, Bruce, 206–208, 209–210
Henry, Patrick, 149
Hess, Henry, 209
heuristics, defined, 124
Heyden, Gaspar van der, 36
Hinks, John, 58
History of the Life and Voyages of Christopher Columbus (Irving), 5
homeomorphisms, 93, 94
homolographic projections, 33
horizontal stretch, of Mercator projection, 38–39, 41
"How Long Is the Coast of Britain?" (Mandelbrot), 73–74, 80
hub-and-spoke system, great circles and, 30
Humboldt, Alexander von, 18

IBM, Mandelbrot at, 71–72
Ibn Sahl, 214n
Illinois, gerrymandering and the Earmuff District, 159–161
information theory, metric space and, 105–108
InSight (Mars rover), investigation into marsquakes and, 225–227
integrals, 43
international waters, determining, 86–89
Islamic Center of Washington, 56–57

Jefferson, Thomas, 67–68
Johnson, Lyndon B., 159
JOIDES Resolution, 211
Jones, Wilfred, 161–162

Kempe, Alfred, 137–138, 139
Kennedy, J. P., 144–145
Koch, Helge von, 75
Koch, John, 141

Koch, Tom, 184–185
Koch snowflake / Koch curve, 75–76, 76f
Koelemeijer, Paula, 220–221, 222–223
König, Dénes, 121
Königberg's seven bridges problem, 114–121, 116f, 117f, 120f, 121
Korean Air Lines Flight 007, 62
Kremer, Geert de. *See* Mercator, Gerardus
Kühn, Heinrich, 115

Lambert, Johann Heinrich, 48
Lambeth Waterworks Company, cholera and, 174–176, 175f
Lamont Geological Observatory (Lamont-Doherty Earth Observatory), 206, 208, 209–210, 219–220
Larcom, Shaun, 94
large low-velocity provinces (LLVPs), 223
latitude, 2f; measuring degree of, 15–17; Mercator projection and, 39, 41–42
Legible London, 98–99, 100f
Lehmann, Inge, 216–220, 223, 230
lidar (light detection and ranging) technology, autonomous driving and, 229
Livingstone, Ken, 98
logarithms, understanding Mercator projection and, 43
London Tube map, 90–92, 91f, 93, 94–95, 230
longitude, 2f; Mercator projection and, 37–38, 41
Lopez, Raymond, 190
loxodrome, 37

Madison, James, 149
magnetic north, 40n, 57
Mandelbrojt, Szalom, 71
Mandelbrot, Benoit B., 71–72, 77; fractional dimensions and, 79–81; Richardson and, 73, 79–80
Manhattan metric, 103–104
mantle, composition of Earth's, 222–223
mantle plumes (blobs), 223
mapmaking: deciding what to omit, 66–67; distortion and, 25–29; Mercator projection, 36–40; projections, 33–36; Rand McNally world map, 53–56; Remarkable Theorem and, 25
maps, autonomous cars and, 228–230. *See also* projections
Markov chain Monte Carlo (MCMC) algorithm, 157–158
Markovich, Jeremy, 147, 148
Mars, investigation of marsquakes and interior, 225–227

Maryland, examination of electoral map for gerrymandering, 158–159
McGhee, Eric, 151–153
Meals on Wheels algorithm, 126–128, 127f
measurement: of borders, 69–71; of curves, 75–78
measurement error, 71
Measure of the Earth (Ferreiro), 14
Mecca, maps and determining orientation to, 56–58
medical mapping, 181–184. *See also* cholera
Menger, Karl, 123
Mercator, Gerardus, 2, 13; creation of Mercator projection, 36–40; Dee and, 46–47
Mercator effect, 50–52
Mercator projection, 2–3; Cold War powers and, 61–62; compared to Gall-Peters projection, 49–50; creation of, 36–40; Eurocentric bias of, 46–47, 51–52; math and, 40–44; North Korean threat and, 60–61; sizes and, 44–47
meridians, 2f; Mercator projection and, 37–38, 41
messenger problem, 123
metrics: cognitive maps and, 110–113; Euclidean, 101–102, 103–104, 110–111; Manhattan, 103–104
metric space, 104–108
Metron, applying Bayesian search to finding Air France Flight 447, 193, 195–198
miasma theory, cholera and, 173–174, 175, 179, 181
minimal counterexample, 138
Monarrez, Tomás, 168, 170
Morison, Samuel Eliot, 11
Morris, Gouverneur, 99–100
Moser, Edvard, 108–109
Moser, May-Britt, 108–109
multibeam sonar, 203–204
multimember districts, gerrymandering and, 164

Nadel, Lynn, 110
Napier, John, 43
National Geographic (magazine), map of ocean floor in, 210
National Geographic Society, 55, 57
NATO phonetic alphabet, 105–106, 107–108
Nature of Maps, The (Robinson and Petchenik), 56
Nautilus (research ship), 204–205
Navajo reservation, effect of gerrymandered districts on, 161–163
navigation: Mercator projection and, 3, 37–40, 47; size of environment and, 112–113; styles of, 112–113; Wright's discussion of mistakes made by sailors in, 40–41

nearest neighbor algorithm, 124–125, 125f
Newton, Isaac, 14
New York City, 99–104; Manhattan metric, 103–104; street plan, 99–101
Nicholson, Francis, 96
nominal scale, 66
North Korean nuclear threat, maps and, 60–61
North Pole, 2f, 25–26, 26f, 34n, 35, 40n, 59, 61
Norway, Fisheries Case and, 87–88
Nunes, Pedro, 37

Obama, Barack, 162–163
observation points, 12, 12f, 17
ocean depth data, Tharp mapping and interpreting, 206–208
Ocean Exploration Trust (OET), 204–205
ocean floor mapping, 200–201, 227; aboard the *Nautilus,* 204–205; discovery of rift valley, 207–208; evidence for continental drift theory and, 207–211; measuring depth, 202–203; Tharp and, 205–211; using multibeam sonar, 203–204
O'Keefe, John, 108–109, 110
Oldham, Richard Dixon, 213, 218
On-Road Integrated Optimization and Navigation (ORION) project, 130–131
On the Heavens (Aristotle), 5–6
On the Measure of the Earth (Eratosthenes), 6
On the Mode of Communication of Cholera (Snow), 174, 175f
Orthogonal Map of the World (Peters), 48–49
orthomorphic projections, 33
Osipovitch, Gennadi, 62

"P′" (Lehmann), 219
packing, 146–147
parallels, 2f; Mercator projection and, 37–38, 40
Perelman, Grigori, 140
Peru, Geodesic Mission to the Equator and, 16–17
Petchenik, Barbara Bartz, 56
Peters, Arno, map of, 47–49
Peutinger Map, 97–98
Phélypeaux, Jean-Frédéric Philippe (comte de Maurepas), 15–16
Philosophiæ Naturalis Principia Mathematica (Newton), 14
physiographic maps, 208
Pixar, 82–83
pizza, Gaussian curvature and, 23–24, 24f
place cells, 109
planar graphs: Kempe's four-color theorem and, 137; vertices and, 138n, 139–130
plane wreckage, using Bayesian search to locate, 193–198
plate tectonics, 207
Poincaré conjecture, 140
Poland, attempt at gerrymandering in, 149–150
politics, choice of projection and, 61–62, 63. *See also* gerrymandering
Polo, Marco, 9–10
Polsby–Popper test, 151
Porter, Irwin S., 57
prime meridian, 2f
Pringles, Remarkable Theorem and, 25
probability theory, 142n
projections, 33–36; azimuthal, 35, 35f; conformal or orthomorphic, 33–34; conic, 34–35, 35f; cylindrical, 34, 35f; equidistance, 33–34; homolographic, 33; scale and, 64–65
proof: debate on what a mathematical proof is, 142–143; four-color theorem, 141–143
proof by contradiction, 138
P (primary or pressure) waves, 213, 213f, 224, 226
P wave shadow zone, 215–216, 218

qibla, 57–58
quadrant, 16
quaternions, 136n

Rahim, Mohammed Kamil Abdul, 56–57
RAND Corporation, 123–124
Randel, John, 100
Rand McNally map, aesthetics of, 53–56
random seed and grow algorithm, 157
Rasmussen, Jens Eilstrup, 53
Rasmussen, Lars, 53
ReCom algorithm, 158
Remarkable Theorem, 3, 22–25
RenderMan, 82–83
retroazimuthal projections, 58–59, 59f, 60f
rhumb line, 37, 38f, 58
Richardson, Lewis Fry: on determining country borders, 69–71; Mandelbrot and, 73, 79–80; measuring geographical curves, 76–77, 77f
Rickey, V. Frederick, 41, 42f
rift valley, Tharp and oceanic, 207–208, 209
Robinson, Arthur H., Rand McNally world map and, 53–56
Robinson, Julia, 123–124
Roman Empire, Peutinger Map of road network of, 97–98
Rossmo, Kim, geographic profiling and, 185–189, 190
Rossmo's formula, 188

route scheduling: traveling salesman problem, 122–124; UPS, 121–122, 129–131
Rutherfurd, John, 99–100

salamander, gerrymandering and, 148
satellite-based mapping, 200–201, 202, 211
scale: Alaska boundaries, 83–86; borders and rivers, 67–71, 73; cauliflowers, 71–72; coastline paradox, 70–71, 86–89; conflict between countries and, 69; defined, 65; fractional dimension, 78–81; Mandelbrot and, 73–74, 78–81; measuring curves, 75–78; nominal, 66; power of, 89; representation of on maps, 65, 66; use of fractals, 81–83
Schmidt, Christopher, 53
Schmitt, Otto, 128–129
schools: equalizing gaps among, 170–172; gerrymandered districts, 164–172; measuring segregation, 168–169; in segregated communities, 166–167, 170
Scoresby Sound, 217–218
Scott, Natacha, 49
Seabed 2030 initiative, 201, 204
Seaman, Valentine, 181
Search and Rescue Optimal Planning System (SAROPS), 193
seating plan, 131–133, 132f
secants, 44
Segregation Contribution Index (SCI), 168–169
Seismic Experiment for Interior Structure (SEIS), 225–227
seismic tomography, 221–223, 222f
seismic waves, 213, 227; bending of, 214, 215f; in Earth's core, 215–216, 218–219. *See also* P (primary or pressure) waves; S (secondary or shear) waves
seismometers / seismographs, 212
self-similarity, fractality and, 73
serial killers, geographic profiling and, 186–187
shape: of the Earth, 5–7, 14–18, 30; maps and preservation of, 33
Simmons, Bobby Ray, Jr. "B.o.B.," 30
simultaneous localization and mapping (SLAM), 229–230
sine, 43
SIR models, 182–183
size, sacrificed on Mercator projection, 44–47, 45f
size of environment, navigation and, 112–113
Smith, Alvy Ray, 82
Snell, Willebrord van Royen, 13, 214n
Snell's law, 214, 219
Snow, John, 173–179, 175f, 178f; critics of, 180–181
sonar, 202–203; multibeam, 203–204
sound, 202n
soundings, 202
Sousa, Martim Afonso de, 37
Southwark & Vauxhall (S&V) water company, cholera and, 174–176, 175f
Soviet Union, power of Mercator projection during Cold War and, 61–62
Sowers, Derek, 204–205
space-filling curve, Meals on Wheels algorithm and, 126–127, 130
sphere, geometry of, 25–27, 26f
Spiers, Hugo, 109
stadium (measurement), 8
Stähler, Simon, 227
Statistics of Deadly Quarrels (Richardson), 69
Steinhaus, Hugo, 80–81
Stephanopoulos, Nicholas, 151–153, 161
Sterling, Colleen, 193–195, 196, 198
Stoa, Ryan, 86, 88, 89
Stone, Lawrence, 194
straight-line distance, 101
strait, 202n
subjective probabilities, Bayesian search and, 191–192
Supreme Court, gerrymandering and, 147–148, 152–153, 158n, 159–160
surface waves, 213
S (secondary or shear) waves, 213, 213f, 224, 226
S wave shadow zone, 215–216, 216f
Sylvester, James Joseph, 121n

"Tabula Peutingeriana," 97–98
tangent, 43
taxicab distance, 103n
territorial sea, 87, 88f
Tharp, Marie, 205–208, 210–211, 230
Tietze, Heinrich, 135f
toise, 17
Tolman, Edward C., 109–110
tomography, seismic, 221–223, 222f
topological maps: Catawba Deerskin Map, 96–97, 97f; Legible London, 98–99, 100f; London Tube map, 90–92, 91f, 93, 94–95; Peutinger Map, 97–98; transportation systems and, 94
topology, 92–94, 111, 120
Toscanelli, Paolo dal Pozzo (Paul the Physician), 9–10
traveling salesman problem (TSP), 123–124
triangle inequality, 102, 102f, 104, 110
triangulation, 11–14, 17, 27
triangulation network, 12–13, 13f

trigonometry, 12, 27, 43, 218–219
trilateration, 224
Trump, Donald, 163
Tuchinsky, Philip, 41, 42f
typhoid fever, water and transmission of, 180–181
Tyson, Neil deGrasse, 30

UN Convention on the Law of the Sea, 87, 88f
underground nuclear explosions, tracking, 219
underwater objects, detecting, 202
UN Geneva Conference on the Law of the Sea, 86–87
United Parcel Service (UPS), 121–122, 129–131
United States: Alaskan borders and, 83–86; Catawba Deerskin Map, 96–97, 97f; Four Corners, 133, 134f; legal implications of coastline paradox and, 88–89; projections, Cold War, and, 61–62; western state borders, 67–68. *See also* gerrymandering
Utah, Native Americans and gerrymandering in, 161–163

Velarde, Lorie, 189–190
vertex (vertices): with five neighbors, 138–139, 139f; graphs and, 118–119, 118f, 121; planar graphs and, 138n, 139–140; seating planning and, 131–132, 132f; simplifying maps and, 117–119
Voronoi diagrams, 178–179, 183–184, 184f
Voronoy, Georgy Feodosevich, 178
voting discrimination, gerrymandering and, 159–161, 162

walking distance, 98–99, 100f, 101
Warren, William H., 111
water, transmission of disease and, 174–181
water depth, measuring, 202–203
Waymo, 228–230
Web Mercator, 52–53
Wegener, Alfred, 210
Whitehead, Henry, 179–180
Wilson, Robin, 138
worldviews, effect of projections on, 50–52, 63
Wright, Edward, 40–42, 44